Tribology and Surface Engineering

Tribology and Surface Engineering

Special Issue Editor

Aleksander Lisiecki

MDPI • Basel • Beijing • Wuhan • Barcelona • Belgrade

MDPI

Special Issue Editor
Aleksander Lisiecki
Silesian University
of Technology
Poland

Editorial Office
MDPI
St. Alban-Anlage 66
4052 Basel, Switzerland

This is a reprint of articles from the Special Issue published online in the open access journal *Coatings* (ISSN 2079-6412) from 2018 to 2019 (available at: https://www.mdpi.com/journal/coatings/special_issues/tribol_surf_eng).

For citation purposes, cite each article independently as indicated on the article page online and as indicated below:

LastName, A.A.; LastName, B.B.; LastName, C.C. Article Title. *Journal Name* **Year**, *Article Number*, Page Range.

ISBN 978-3-03928-084-1 (Pbk)
ISBN 978-3-03928-085-8 (PDF)

Cover image courtesy of Aleksander Lisiecki.

Contents

About the Special Issue Editor

Aleksander Lisiecki is Professor at Silesian University of Technology in Gliwice, Poland, where he has been working since graduating in 1998. He completed two internships at the University of Warwick, Warwick Manufacturing Centre in Coventry, United Kingdom, within the framework of the international program TEMPUS in 1999. He obtained his PhD in the discipline of materials science in 2001 and next the habilitation degree in 2017. He has served as Vice-Head of the Department of Welding Engineering since his appointment in 2017 and is involved in both teaching and research. He has executed many scientific projects and been involved in numerous research and development projects in the field of manufacturing technologies, especially welding, laser processing, and coatings. He is the author or coauthor of over 200 publications in Polish and English, 11 patents, and 18 patent applications. His main areas of research and interests include welding technologies, additive manufacturing of metal components, joining of metals and alloys, laser materials processing, and wear-protective coatings.

Preface to "Tribology and Surface Engineering"

The term *"tribology"* comes from a combination of the Greek words *"τρίβω"* (*tribo; means "rubbing" in English*) and *"λόγια"* (*logia; means "knowledge, study" in English*), which can be translated as "study on friction". The term "tribology" was used first time by Peter Jost in 1966, in a report on in which he noted the importance of costs of friction, wear, and corrosion to the UK economy. However, the phenomena concerned by tribology have been known since the beginning of human history. Since the first use of the term tribology, it has acquired importance and is widely used. Tribology can be defined as an interdisciplinary area of science and technology focused on interacting surfaces in relative motion, including the study of friction, wear, lubricants, as well as the design of bearings. The interdisciplinary science of tribology covers such engineering subject as chemistry, mathematics, physics, material science, solid and fluid mechanics, and heat transfer.

Developments in the field of tribology are associated with the development of the world economy and the desire to increase the efficiency and durability of machines, devices, and tools as well as to increase the productivity of technological processes and to reduce energy consumption and emissions. Tribology is of great practical importance because the correct operation and functioning of many mechanical, electrochemical, biomedical, or even biological systems are dependent on the corresponding coefficient of friction (low or high), and also on the controlled degree of wear. Examples illustrating the above systems may be mechanical components and devices such gears, transmissions, plain and rolling bearings, engines, tires, biomedical components such artificial joints and implants, material processing technologies such cutting, machining, grinding, polishing, or electrical contacts in electrotechnical devices. The significance of the problem may be evidenced by the fact that according to recent estimates, the loss of natural resources as a direct cause of friction and wear exceeds 6

Another area of science and technology directly related to tribology is the "surface engineering". Surface engineering, as a subdiscipline of materials science, was established relatively recently, and the first Surface Engineering Research Institute was founded at Birmingham University of Technology in 1983. Surface engineering deals with designing and shaping the functional surface properties of components such physical, chemical, electrical, magnetic or mechanical, which are different from that of the bulk material. Surface properties are shaped to ensure appropriate working conditions or the required durability under specific loading conditions and environments, e.g., enhanced corrosion and wear resistance, oxidation resistance at elevated temperatures, enhanced absorptivity or reflectivity, biocompatibility, wetting, etc. Surface engineering also applies to the shaping of aesthetic characteristics of components (decorative elements). The required distinct properties of surface and subsurface regions are provided in thin layers ranging from a few nanometers to tens of millimeters.

In general, surface modification methods used in surface engineering can be divided into three basic groups. The first is associated with microstructural modification in the near-surface region (e.g., heat treatment, mechanical deformation, or surface melting). The second involves chemical composition changes of surface and near-surface region (e.g., thermochemical treatment, alloying, ion implantation). The third group consists of developing surface layers and overlays (e.g., coatings, plating, deposition). Surface engineering also includes the methods of investigation and assessment the quality (e.g., presence of discontinuities such as porosity, cracks, and inclusions) and properties (e.g., hardness, corrosion resistance, or wear resistance) of modified surfaces and coatings. Nanosurface engineering has also recently emerged as a result of the development in materials science, technology, and manufacturing, allowing the production of surface nanofilms or nanostructured coatings.

The articles in this book collection are all related to the very important scientific and technological issues detailed above.

Aleksander Lisiecki
Special Issue Editor

coatings

MDPI

Editorial

Tribology and Surface Engineering

Aleksander Lisiecki

Department of Welding Engineering, Faculty of Mechanical Engineering, Silesian University of Technology, Konarskiego 18A street, 44-100 Gliwice, Poland; aleksander.lisiecki@polsl.pl

Received: 2 October 2019; Accepted: 11 October 2019; Published: 13 October 2019

Abstract: The Special Issue on Tribology and Surface Engineering includes nine research articles and one review article. It concerns a very important problem of resistance to wear and shaping the properties of the surface layers of different materials by different methods and technologies. The topics of the presented research articles include reactive direct current magnetron sputtering of silicon nitrides on implants, laser surface modification of aeroengine turbine blades, laser micro-texturing of titanium alloy to increase the tribological characteristics, electroplating of Cu–Sn composite coatings incorporated with Polytetrafluoroethylene (PTFE) and TiO_2 particles, arc spraying of self-lubricous coatings, high velocity oxygen fuel (HVOF) spraying and gas nitriding of stainless steel coatings, HVOF spraying composite WC-Co coatings, testing of coatings deposited by physical vapour deposition (PVD), and also analysis of material removal and surface creation in wood sanding. The special issue provides valuable knowledge based on theoretical and empirical study in the field of coating technologies, as well as characterization of coatings, and wear phenomena.

Keywords: surface engineering; coatings; tribology; wear resistance; properties of surface layers

1. Introduction

The progressive wear of moving parts and components of machines or tools under operating conditions is a natural phenomenon, leading to a gradual decrease in performance, efficiency and work parameters of a given element. The used machine parts or tools need to be replaced with new ones, which is usually associated with the need to temporarily shut down the machine or the entire technological process.

Meanwhile, global industry constantly strives to increase the efficiency and effectiveness of technological processes, as well as machines and vehicles. This makes the working conditions of tools and machine parts increasingly difficult. Therefore, the industry is constantly searching for new and advanced structural and tool materials, as well as methods of their production, which in specific operating conditions will ensure maximum durability, at an acceptable level of unit costs.

One of the most significant and interesting achievements in materials engineering and manufacturing processes is the development and introduction of composite materials to the industry. Research and development on production methods of metal matrix composites (MMC) are carried out in many research centers around the world [1–9]. These types of composite materials combine the advantages of the metal matrix characterized by high ductility with the reinforcing material. Therefore, such composite materials have excellent fatigue strength, high ultimate strength, and simultaneously low specific weight. In the case of composite coatings, usually dispersion particles or phases with high hardness provide increased wear resistance.

Another significant achievement in materials engineering and manufacturing processes in recent years is the introduction and now widespread use of nanomaterials, as well as amorphous materials [10–16]. However, intensive research is still ongoing, and the attention of researchers from around the world is focused on manufacturing processes and the application of nanomaterials due to the high potential of this group of materials, characterized by grain size on a nanometric scale, most often in the range below 100 nm. Nanostructured materials—both metallic, and composite—have functional properties (e.g., mechanical strength) often many times higher than conventional materials. Even by grain refinement and creating a nanostructure in the case of already known and conventional engineering materials, it is possible to significantly improve physical, chemical, and mechanical properties—such as e.g., increase in hardness and strength—increase in wear resistance of contact surfaces, increase in ductility of fragile materials, etc. For example, the tensile strength of copper with grains of the order of 50 microns does not exceed 500 MPa, while after grains refinement to dimensions of the order of 8 nm, the strength increases up to five times to 2500 MPa.

The manufacturing of coatings of modern composite and metallic materials in micro or nanometric scale, as well as the manufacturing of surface layers with special properties is possible only by applying the most advanced technologies of coatings and surface engineering.

Currently, the most commonly-used technologies of coatings in industry are as follows:

- Arc and plasma cladding technologies ensuring high melting efficiency of the additive material (consumable). However, the disadvantage is the relatively huge amounts of heat introduced into the substrate material make it difficult to obtain fine-grained structures. For this reason, these technologies are used usually for cladding of relatively large and massive components.
- Laser beam and electron beam cladding, providing very low heat input, low dilution, and narrow heat affected zone.
- Thermal spraying technologies including flame, arc, and plasma spraying, high velocity oxygen fuel spraying (HVOF), and cold spraying (CS) at supersonic gas jet velocities, that allow the production of coatings with a structure and chemical composition unattainable by the cladding technologies. The disadvantage is that the coatings are characterized by discontinuity and often considerable porosity, which reduces their utility values and limits the scope of practical application.
- PVD and chemical vapor deposition (CVD) coating technologies that allow the creation of metallic and composite nanostructured coatings. However, in this case the thickness of the coatings is small and the manufacturing process is time consuming, and also extremely expensive.

Another issue is the shaping properties of surface layers to improve the tribological characteristic and wear resistance by various surface engineering methods including heat treatment and thermochemical treatment.

Various methods of surface heat treatment to improve the tribological properties of metals and alloys were investigated and are widely described in the literature [17–21]. The most interesting and most advanced are laser and plasma surface treatment, and ion implantation. The basic principle of the surface heat treatment methods is to change the microstructure in the surface layer, without changing the chemical composition, by extremely high heating and cooling rates. Depending on the processing parameters and conditions the mentioned above methods can be also used for thermochemical treatment. Thanks to the thermochemical surface treatment coupled with enrichment of the surface layer by different elements, the corrosion resistance and wear resistance can be improved, as well as the friction coefficient can be significantly lowered. Oxidation, carburizing, and nitriding are the most popular and widely used methods of thermochemical surface treatment of metals and alloys, in particular in the case of titanium and titanium alloys.

Summarizing, the intensity and the type of wear are dependent on the type and intensity of the load, and also on the environmental factors such as corrosive liquid or gaseous agents, temperature, etc. Therefore, in order to provide the durability and reliability of components of machines or tools an individual approach to each of the working surfaces is required, choosing the proper material, design its microstructure, and application the optimal method of manufacturing or processing.

2. This Special Issue

This special issue, entitled "Tribology and Surface Engineering", is a complementary and valuable resource of knowledge in the field of phenomena and mechanisms of surface wear, and methods of enhancing tribological properties of working surfaces.

Liu et al. [22] provide comprehensive review of the analytical and empirical studies on micro pitting phenomena in the case of steel gears used in wind turbines, helicopters, or ships. They identified and pointed several relevant factors influencing the micro pitting behaviors, e.g., gear materials, surface topographies, lubrication properties, working conditions. They also described mechanisms of wear, and indicated the way to improve the micro pitting resistance of the gears. The information presented in the review can be very helpful and useful for designers of modern heavy-load, high-speed mechanisms.

In turn, a very similar problem related to gear wear was investigated and described by Michalczewski et al. [23]. They pointed out that the type of the oil used for the transmission can have a significant impact on its durability and reliability. Therefore, they investigated the influence of three commercially available industrial gear oils on test samples with a new type of W-DLC/CrN coating. One mineral oil, and two synthetic oils with polyalphaolefin (PAO) and polyalkylene glycol (PAG) bases, respectively, were applied in the study. Based on the abrasion, scuffing, and pitting tests, they showed that synthetic polyalphaolefin (PAO) type oil provides the most favorable working conditions and the highest durability.

The microstructure and tribological properties of metallic coatings produced by high velocity oxygen fuel spraying (HVOF) were investigated by Lindner et al. [24], while the composite WC-Co type coatings produced by HVOF spraying were investigated by Ding et al. [25]. Lindner et al. applied additional thermochemical treatment of the HVOF-sprayed AISI 316L coatings to improve the wear resistance. The gas nitriding process was conducted at different temperatures. They successfully enriched the coatings in as-sprayed conditions by the nitrogen, and showed significant increase in wear resistance. In turn, Ding et al. produced composite conventional, multimodal, and nanostructured WC-12Co coatings with different WC sizes and distributions were prepared by HVOF spraying. The microstructure, phase composition, hardness, porosity, and cavitation erosion were investigated. The results revealed that the despite serious decarburization of nanostructured WC-Co coatings resulting in formation of W_2C and W phases, the nanostructured WC-Co coatings have the densest microstructure with lowest porosity, the highest fracture toughness, and also they exhibit the highest resistance to cavitation erosion wear.

The next articles concern the impact of reducing the coefficient of friction between contact surfaces on wear phenomena. Tillmann et al. [26] proposed a method for creating self-lubricous coatings based on arc spraying of vanadium containing iron-based deposit. The wear characteristic was investigated under dry sliding experiments, while the worn surfaces were examined by means of electron microscopy and energy dispersive X-ray (EDX) spectroscopy. They found that the vanadium-containing coatings exhibited a distinctly reduction of the coefficient of friction above 450 °C temperature, due to prevalence of specific vanadium oxides which promote a self-lubricating ability of the coating. Silicon nitride (SiN_x) coatings are considered as bearing surfaces for joint implants, due to their low wear rate and the good biocompatibility. Therefore, reactive direct current magnetron sputtering was applied by Filho et al. [27] to coat the CoCrMo disc samples with a CrN interlayer, followed by a SiN_x top layer, which was deposited by reactive high-power impulse magnetron sputtering. The phase composition, surface roughness, hardness distribution, and wear rate of the test coatings were investigated. They found that the bias voltages have a significant influence on the performance of SiN_x coatings, characterized by low wear rates. The promising results of study, conducted by Filho et al., support further development of silicon nitride-based coatings towards clinical application. In turn, Ying et al. [28] investigated the effect of TiO_2 particles and PTFE emulsion on properties of Cu–Sn composite coatings. Cu–Sn, Cu–Sn–TiO_2, Cu–Sn–PTFE, and Cu–Sn–PTFE–TiO_2 coatings type were electroplated with a pulsed power supply. The microstructure, phase composition, microhardness, corrosion resistance, and tribological properties were investigated. They described the influence of PTFE and TiO_2 on the microstructure, corrosion resistance, hardness, and tribological properties of the test coatings. They proved that presence of both PTFE and TiO_2 in the deposited coating leads to a lower friction coefficient of 0.1 and higher wear and corrosion resistance. Vazquez Martinez et al. [29] applied laser micro-texturing of the Ti6Al4V alloy surface to improve the friction, wear, and wettability behavior under sliding conditions. They investigated the influence of processing parameters such scanning speed of the laser beam, and the energy density of pulse on the tribological characteristic of the titanium alloy, including measurements of the contact angle using water as a contact fluid. The wear mechanisms were also studied and determined by means of microscopic observations. They found a strong dependence between the wear behavior and the laser patterning parameters. The micro-texturing of the surface caused reduction in wear up to a 70%, compared to untreated surfaces of Ti6Al4V alloy.

Another example of laser beam application in surface engineering for repair cladding and shaping properties of surface layers was demonstrated by Liu et al. [30]. Authors investigated the laser cladding of K417G Ni-based superalloy by analyzing the possibility of built-up cladding of worn turbine blades. Additionally, the laser surface remelting was applied for controlling the cracking sensitivity. Microstructure, hardness, and tribological properties of the base metal, coating after cladding, and after additional remelting were determined and described in details. Authors showed that, despite high tendency to cracking of the investigated Ni-based superalloy, the additional laser remelting process is advantageous, because it results in decreasing the size of cracks of the multilayer laser clads.

The last research article described here concerns investigations on sanding process of medium-density fiberboard (MDF) and Korean Pine. Zhang et al. [31] determined the mechanisms of material removal and the influence of processing parameters on the surface quality of the investigated materials. Authors declare that the finding and used approaches could provide insights to investigate other wood species or wood composites to improve the efficiency of sanding and simultaneously the surface quality.

3. Concluding Remarks

The special issue was very successful due to valuable articles submitted, a wide range of research problems undertaken, as well as in-depth analysis of the state of the art. The Special Issue consists of 10 papers; however, the total number of manuscripts submitted to the Special Issue was almost twice that. Unfortunately, some of the manuscripts have not gone through a very rigorous review process. Such a large interest and a large number of articles show the importance of the issue and themes.

It should be noted that in the field of materials engineering and surface engineering, each original scientific and research article is the culmination of hard research work and long-term study, usually a team of scientists. In turn, the results of such research have practical significance and contribute to the development of civilization.

Thanks to the MDPI Publisher and the reputable Open Access Journal "Coatings", interesting and innovative achievements of interdisciplinary teams of scientists from different countries can be presented to a wide range of readers around the world.

That is why I strongly encourage readers to thoroughly read all articles, and also I encourage the scientists to continue their research work and publish interesting results.

Funding: This publication received no external funding.

Acknowledgments: As the Editor of the Special Issue, I would like to thank all the authors of the submitted articles, as well as the reviewers, editors, and all who contributed to publishing the Special Issue.

Conflicts of Interest: The author declares no conflict of interest.

References

1. Kamat, A.M.; Copley, S.M.; Segall, A.E.; Todd, J.A. Laser-sustained plasma (LSP) nitriding of titanium: A review. *Coatings* **2019**, *9*, 283. [CrossRef]

2.   Kusinski, J.; Kąc, S.; Kopia, A.; Radziszewska, A.; Rozmus-Górnikowska, M.; Major, B.; Major, L.; Marczak, J.; Lisiecki, A. Laser modification of the materials surface layer—A review paper. *Bull. Pol. Acad. Sci. Tech. Sci.* **2012**, *60*, 711–728. [CrossRef]

3.   Janicki, D. Microstructure and sliding wear behaviour of in-situ TiC-reinforced composite surface layers fabricated on ductile cast iron by laser alloying. *Materials* **2018**, *11*, 75. [CrossRef]

4.   Lisiecki, A. Mechanisms of hardness increase for composite surface layers during laser gas nitriding of the Ti6A14V alloy. *Mater. Technol.* **2017**, *51*, 577–583.

5.   Lisiecki, A.; Kurc-Lisiecka, A. Erosion wear resistance of titanium-matrix composite Ti/TiN produced by diode-laser gas nitriding. *Mater. Tehnol.* **2017**, *51*, 29–34. [CrossRef]

6.   Lisiecki, A. Comparison of titanium metal matrix composite surface layers produced during laser gas nitriding of Ti6Al4V alloy by different types of lasers. *Arch. Met. Mater.* **2016**, *61*, 1777–1784. [CrossRef]

7.   Lisiecki, A. Titanium matrix composite Ti/TiN produced by diode laser gas nitriding. *Metals* **2015**, *5*, 54–69. [CrossRef]

8.   Tański, T.; Matysiak, W. Synthesis of the novel type of bimodal ceramic nanowires from polymer and composite fibrous mats. *Nanomaterials* **2018**, *8*, 179. [CrossRef]

9.   Kurc-Lisiecka, A.; Lisiecki, A. Laser welding of new grade of advanced high strength steel Domex 960. *Mater. Technol.* **2017**, *51*, 199–204.

10.   Pilarczyk, W. Structure and properties of Zr-based bulk metallic glasses in as-cast state and after laser welding. *Materials* **2018**, *11*, 1117. [CrossRef]

11.   Kik, T.; Górka, J. Numerical simulations of laser and hybrid S700MC T-joint welding. *Materials* **2019**, *12*, 516. [CrossRef] [PubMed]

12.   Tomków, J.; Rogalski, G.; Fydrych, D.; Łabanowski, J. Advantages of the application of the temper bead welding technique during wet welding. *Materials* **2019**, *12*, 915. [CrossRef] [PubMed]

13.   Kaźmierczak-Bałata, A.; Mazur, J. Effect of carbon nanoparticle reinforcement on mechanical and thermal properties of silicon carbide ceramics. *Ceram. Int.* **2018**, *44*, 10273–10280. [CrossRef]

14.   Górka, J.; Czupryński, A.; Żuk, M.; Adamiak, M.; Kopyść, A. Properties and structure of deposited nanocrystalline coatings in relation to selected construction materials resistant to abrasive wear. *Materials* **2018**, *11*, 1184. [CrossRef]

15.   Kurc-Lisiecka, A. Impact toughness of laser-welded butt joints of the new steel grade Strenx 1100MC. *Mater. Technol.* **2017**, *51*, 643–649.

16.   Haghighi, M.; Shaeri, M.H.; Sedghi, A.; Djavanroodi, F. Effect of graphene nanosheets content on microstructure and mechanical properties of titanium matrix composite produced by cold pressing and sintering. *Nanomaterials* **2018**, *8*, 1024. [CrossRef]

17.   Lukaszkowicz, K.; Jonda, E.; Sondor, J.; Balin, K.; Kubacki, J.M. Characteristics of the AlTiCrN + DLC coating deposited with a cathodic arc and the PACVD process. *Mater. Tehnol.* **2016**, *50*, 175–181. [CrossRef]

18.   Bonek, M. The investigation of microstructures and properties of high speed steel HS6-5-2-5 after laser alloying. *Arch. Met. Mater.* **2014**, *59*, 1647–1651. [CrossRef]

19.   Lisiecki, A. Study of optical properties of surface layers produced by laser surface melting and laser surface nitriding of titanium alloy. *Materials* **2019**, *12*, 3112. [CrossRef]

20.   Lisiecki, A.; Piwnik, J. Tribological characteristic of titanium alloy surface layers produced by diode laser gas nitriding. *Arch. Met. Mater.* **2016**, *61*, 543–552. [CrossRef]

21.   Janicki, D. Fabrication of high chromium white iron surface layers on ductile cast iron substrate by laser surface alloying. *Stroj. Vestn.* **2017**, *63*, 705–714. [CrossRef]

22.   Liu, H.; Liu, H.; Zhu, C.; Zhou, Y. A review on micropitting studies of steel gears. *Coatings* **2019**, *9*, 42. [CrossRef]

23.   Michalczewski, R.; Kalbarczyk, M.; Mańkowska-Snopczyńska, A.; Osuch-Słomka, E.; Piekoszewski, W.; Snarski-Adamski, A.; Szczerek, M.; Tuszyński, W.; Wulczyński, J.; Wieczorek, A. The effect of a gear oil on abrasion, scuffing, and pitting of the DLC-coated 18CrNiMo7-6 steel. *Coatings* **2019**, *9*, 2. [CrossRef]

24.   Lindner, T.; Kutschmann, P.; Löbel, M.; Lampke, T. Hardening of HVOF-sprayed austenitic stainless-steel coatings by gas nitriding. *Coatings* **2018**, *8*, 348. [CrossRef]

25.   Ding, X.; Ke, D.; Yuan, C.; Ding, Z.; Cheng, X. Microstructure and cavitation erosion resistance of HVOF deposited WC-Co coatings with different sized WC. *Coatings* **2018**, *8*, 307. [CrossRef]

26. Tillmann, W.; Hagen, L.; Kokalj, D.; Paulus, M.; Tolan, M. Temperature-induced formation of lubricous oxides in vanadium containing iron-based arc sprayed coatings. *Coatings* **2019**, *9*, 18. [CrossRef]
27. Filho, L.; Schmidt, S.; Leifer, K.; Engqvist, H.; Högberg, H.; Persson, C. Towards functional silicon nitride coatings for joint replacements. *Coatings* **2019**, *9*, 73. [CrossRef]
28. Ying, L.; Fu, Z.; Wu, K.; Wu, C.; Zhu, T.; Xie, Y.; Wang, G. Effect of TiO_2 sol and PTFE emulsion on properties of Cu–Sn antiwear and friction reduction coatings. *Coatings* **2019**, *9*, 59. [CrossRef]
29. Vazquez Martinez, J.M.; Del Sol Illana, I.; Iglesias Victoria, P.; Salguero, J. Assessment the sliding wear behavior of laser microtexturing Ti6Al4V under wet conditions. *Coatings* **2019**, *9*, 67. [CrossRef]
30. Liu, S.; Yu, H.; Wang, Y.; Zhang, X.; Li, J.; Chen, S.; Liu, C. Cracking, microstructure and tribological properties of laser formed and remelted K417G Ni-based superalloy. *Coatings* **2019**, *9*, 71. [CrossRef]
31. Zhang, J.; Ying, J.; Cheng, F.; Liu, H.; Luo, B.; Li, L. Investigating the sanding process of medium-density fiberboard and Korean Pine for material removal and surface creation. *Coatings* **2018**, *8*, 416. [CrossRef]

coatings

MDPI

Review

A Review on Micropitting Studies of Steel Gears

Huaiju Liu *, Heli Liu, Caichao Zhu and Ye Zhou

State Key Laboratory of Mechanical Transmissions, Chongqing University, Chongqing 400030, China; heli_liu@cqu.edu.cn (H.L.); cczhu@cqu.edu.cn (C.Z.); zhouye@cqu.edu.cn (Y.Z.)
* Correspondence: huaijuliu@cqu.edu.cn; Tel.: +86-23-6511-1192

Received: 14 November 2018; Accepted: 9 January 2019; Published: 14 January 2019

Abstract: With the mounting application of carburized or case-hardening gears and higher requirements of heavy-load, high-speed in mechanical systems such as wind turbines, helicopters, ships, etc., contact fatigue issues of gears are becoming more preponderant. Recently, significant improvements have been made on the gear manufacturing process to control subsurface-initiated failures, hence, gear surface-initiated damages, such as micropitting, should be given more attention. The diversity of the influence factors, including gear materials, surface topographies, lubrication properties, working conditions, etc., are necessary to be taken into account when analyzing gear micropitting behaviors. Although remarkable developments in micropitting studies have been achieved recently by many researchers and engineers on both theoretical and experimental fields, large amounts of investigations are yet to be further launched to thoroughly understand the micropitting mechanism. This work reviews recent relevant studies on the micropitting of steel gears, especially the competitive phenomenon that occurs among several contact fatigue failure modes when considering gear tooth surface wear evolution. Meanwhile, the corresponding recent research results about gear micropitting issues obtained by the authors are also displayed for more detailed explanations.

Keywords: micropitting; steel gears; wear; competitive mechanism

---

1. Introduction

Micropitting—also known as peeling, superficial spalling or grey staining—is a surface-initiated fatigue phenomenon occurring during the gear meshing process between interacting surfaces in which a cluster of micropits presents a grey-colored appearance. Compared to classical macropitting which occurs on the nominal Hertzian contact zone and can be identified with the naked eyes, the dimension of the gear micropits can span no more than tens of microns in depth and width along the tooth profile [1], while its length along the axial direction can generally reach 100 μm [2]. As the extensive application of carburized or other case-hardening gears made on the suppression of the subsurface initiate failures, micropitting tends to draw more attention as it can significantly restrict the fatigue lives and reliabilities of the relative mechanical systems [3]. Actually, micropitting is not a instantaneous catastrophic failure mode, but it may lead to vibrations, noises and misalignments, or even contribute to the appearance of other failure modes including pitting, spalling, scuffing and tooth breakage. Höhn et al. [4] stated in 1996 that micropitting had become the most limiting factor of gear behaviors. The National Renewable Energy Laboratory (NREL), Golden, CO, USA, started a gearbox reliability project and hosted a wind turbine micropitting workshop, which was held at the National Wind Technology Center in Boulder, CO, USA, on April 2009 [5]. Just two years later, the NREL reported on the correlation between surface roughness and micropitting at the "Wind Turbine Tribology Seminar", and the effect of the superfinishing technique was emphasized [6]. The malfunctions and accidents on wind turbine gears caused by micropitting in operation are shown in Figure 1.

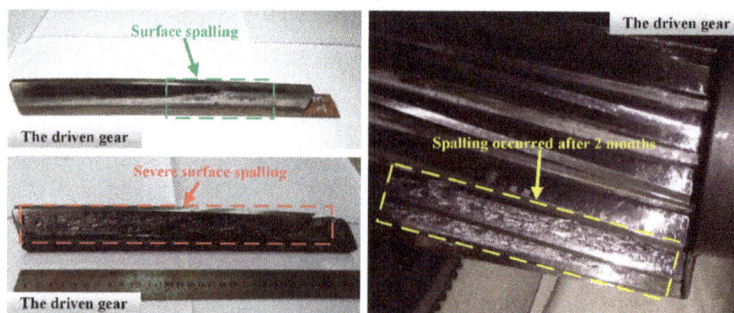

Figure 1. The failures observed in wind turbine gears related to micropitting.

Micropitting often starts during the running-in stage (approximately 10^5 to 10^6 loading cycles) [7]. As revealed in the field operation, micropitting is often observed near the tooth dedendum, whereas there are also cases in which micropits are found in both the addendum and the dedendum (probably including the pitch line or along the tooth tip). Brandão et al. [8] developed a numerical model of surface-initiated failures and then verified it by applying an actual micropitting test on the carburized gears [9]. The results clearly showed that micropitting could be observed between the pitch line region and the tooth root after the loading period. Additionally, Martins et al. [7] also implemented gear micropitting tests to understand the roughness evolution and micropitting initiation. They found that the occurrence position of each micropit could be significantly affected by tooth surface texture characteristics. The micropits could even be generated on the gear surface roughness valleys, which was resulted from the roughness distribution characteristic of the counter surface.

It is well established that different types of gears show distinguished micropitting resistances. Errichello [5] from the NREL, concluded that helical gears are more appropriate for the micropitting tests than spur gears because they have less vibration and a lower transmission error. Besides, helical gears are less sensitive to profile changes and could operate much more smoothly, leading to the better quality of the fatigue test results. The investigation completed by the British Mechanical Power Transmissions Association found that helical gears produced by case-carburized 16MnCr5 or EN36 steels could give a better micropitting resistance than spur gears [10]. Figure 2 displays the reproduced micropitting damages found on gear tooth flanks of both helical and spur gears after the experiments conducted by the Newcastle University, UK.

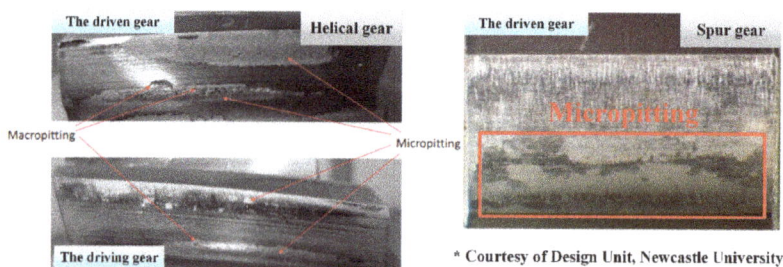

Figure 2. The reproduced micropits observed on the case-hardening and ground gear tooth flank after the experiments.

It is now generally recognized that micropitting is not only a mechanical or physical problem, but also a chemical problem [10]. In principle, within the nominal contact zone, lubricants will chemically react with surfaces and may subsequently generate stress corrosion cracking, etching or a reaction layer [10]. Winter et al. [11] found that the micropitting resistance would be enhanced by using lubricants containing additives based on sulfurated and phosphorous compounds. Brechot et al. [12] stated that the application of anti-wear (AW) and extreme pressure (EP) additives could prevent scuffing or wear but might promote the micropitting process. Sun et al. [13] investigated the tribo-electrochemical behavior of AISI 304 austenitic stainless steel in a 0.5 M NaCl solution at an anodic potential of 70 mV (SCE) under controlled sliding and electrochemical conditions. The results indicated that the micropits formation was determined by the combined effects of contact load, contact frequency and sliding time.

The contact fatigue performance in lubricated Hertzian contacts has been studied since decades ago by Way [14], Dowson [15], and other researchers [16–18]. With the view of quantifying different lubrication conditions, an important relevant parameter was defined as the specific lubricant film thickness λ, namely the lambda ratio, by Tallian [19]:

$$\lambda = \frac{h_0}{R_q} \ (R_q = \sqrt{R_{q1}^2 + R_{q2}^2}) \tag{1}$$

where h_0 means the lubricant film thickness with the assumption that the contact surface is smooth, R_q is called the combined root-mean-square (*RMS*) (certainly, R_q here can be replaced by the effective arithmetic mean roughness value R_a [20]), R_{q1}, R_{q2} can be regarded as the *RMS* values of the two contact tooth surface roughness. Micropitting is remarkably influenced by the specific lubricant film thickness λ [21]. As recorded in international technique report ISO/TR 15144:2010 [20], a safety factor S_λ based on the specific lubricant film thickness λ is also introduced to represent the micropitting load capacity:

$$S_\lambda = \frac{\lambda_{min}}{\lambda_p} \geq S_{\lambda,\ min} \tag{2}$$

where λ_{min} denotes the minimum specific lubricant film thickness in the contact area, λ_p stands for the permissible specific lubricant film thickness, $S_{\lambda,min}$ represents the minimum required safety factor. When $S_\lambda < 1$, a high risk of micropitting is recommended; when $1 < S_\lambda < 2$, a moderate micropitting risk is recommended; when $S_\lambda > 2$, a low micropitting risk is recommended [22,23]. Long et al. [24] from the University of Sheffield, South Yorkshire, UK, carried out an investigation on the influence of load variation on gear tooth surface micropitting based on ISO/TR 15144:2010 [20]. Al-Tubi et al. [25] from the same research group studied the gear micropitting initiation and propagation under varying loading conditions based on the ISO/TR 15144:2010 and the revised version ISO/TR 15144:2014 [26].

Although many investigations have been implemented on the gear micropitting issues, and great progressions have been achieved by many researchers and engineers, the detailed mechanisms of micropitting crack initiation and propagation still remain unclear. The future research and real industrial challenges are still severe until micropitting can be well monitored, detected, prevented, and the micropitting mechanism, together with the correlation with other failure modes on gears, could be further explored. Several aspects of the micropitting analysis, such as the thermal effect, the plastic deformation [27], phase transformation, gear types, etc., also need to be persistently promoted.

For the purpose of further understanding, the gear micropitting mechanism and, consequently, efficiently controlling the occurrence of micropitting in the real gear industry, plenty of studies are reviewed in this work to illustrate the steel gear micropitting phenomenon from various aspects. With the consideration of the wear process, a numerical model incorporating material mechanical properties, residual stress gradient and the measured surface roughness of a wind turbine gear pair were developed under the elastohydrodynamic lubrication (EHL) condition by the authors in order to explore the wear effect on the competitive mechanism between micropitting and pitting during the cycling loading period, which could help lead to a relative in-depth comprehension of micropitting.

2. Some Influence Factors on Gear Micropitting

Based on the above discussions, there are numerous factors that can affect the micropitting phenomena of gear operations. For example, Olia et al. [1] designed a factorial experiment using a twin disc machine for the purpose of assessing the influence of seven factors, namely the gear steel material, lubricant properties, surface finish, load, thermal effect, speed, and slide-roll ratio, on the micropitting initiation and propagation processes. The results could be summarized as follows: the load had the biggest effect on micropitting initiation, and micropitting propagation could be dramatically impacted by the sliding/rolling condition. Therefore, this section summarizes the relative studies on the influence factors of gear micropitting made by the authors and the other researchers.

2.1. Gear Materials and Macroscopic Geometries

As revealed by various experimental observations [28–30], the difference of gear steel materials and the heat treatment process could generate distinguished micropitting resistances. As reported in the Reference [31], high strength austempered ductile iron gears were gradually accepted due to the low production cost, the eventual noise and vibration reduction, and the self-lubricant properties through graphite nodules, resulting in remarkable gear tooth flank capacity. Moreover, the diversity of surface treatments such as carburizing, nitriding, case-hardening, etc. can improve the gear surface fatigue performances by modifying the material properties or forming a hardened layer over the gear substrate.

Wilkinson et al. [32] from the Imperial College London, London, UK, conducted gear tests on different gear materials for micropitting damage. According to the results, carburized AISI M50NiL steel shows no improvement over conventional carburized 4%NiCrMo steel, but the carburized and subsequently nitrided AISI M5ONiL steel could give a substantial improvement. Metallographic examination on a number of micropitted gears was conducted by Olia et al. [33] and they found that the initiation and propagation of cracks leading to the formation of micropits were related to the phase transformations induced by the contact fatigue process. The micropitting propagation rate was obviously larger for the gear steel in which the martensite decay was more pronounced. Le et al. [34] performed experiments on gas nitride gears through a twin-disc machine; the conclusions could be summarized as the fact that despite the equivalent hardness and compressive residual stress gradients, nitrided layers with thin grains and cementite filaments were more resistant to micropitting than coarse microstructure layers. Moreover, Tobie et al. [35] from the Forschungsstelle für Zahnräder und Getriebebau (FZG), Garching bei München, Germany, tried to optimize the carburized gear performance through alloying modification. The conclusion could be summarized as the fact that the gears made of modified 20MnCr5 presented a remarkable lower sensitivity to micropitting than the gears made of modified 20CrMo5. However, more experiments should be implemented on various types of alloying modification techniques to form a universal method. Roy et al. [36] studied the influence of retained austenite (RA) on the micropitting of carburized AISI 8620 steel under the boundary lubrication state. Three levels of RA% were chosen in the tests, namely 0% (low RA), 15% (medium RA) and 70% (high RA), to guarantee the experimental efficiency. With the application of the X-ray diffraction method, the 70% RA sample was proved to present the best micropitting life. This indicated that higher levels of RA% should be guaranteed by appropriate heat treatments, in order to ensure a stable amount RA% remains after the phase transformation for the sake of the micropitting life improvement. However, the effect of RA level on the comprehensive contact fatigue has not been fully revealed.

The gear geometry properties can also have nonnegligible effects on micropitting resistance. In the former work reported in Forschungsvereinigung Antriebstechnik e.V. (FVA, Research Association for Drive Technology) research project 259/I (1994–1998), the influences of gear geometry and gear size on micropitting were discussed in detail. For instance, the diagram of a beveloid gear tooth with transverse profile slope modification and profile crowning modification is schematically illustrated in Figure 3. Several studies concerning the gear geometry or gear tooth modification are presented as follows.

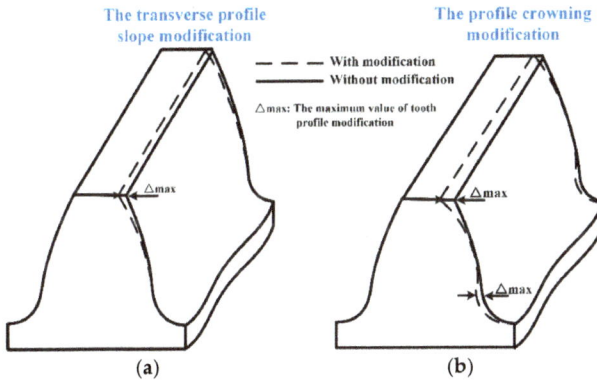

Figure 3. (**a**) The schematic diagram of a beveloid tooth with a transverse profile slope modification and (**b**) profile crowning modification (Reprinted with permission from [37]. Copyright 2018 Elsevier).

Kissling [38] outlined that anti-micropitting capacity could be improved through macrogeometry optimization such as gear tooth profile modifications. Predki et al. [39] studied the effect of gear tooth flank modification on micropitting and concluded that the variation of tip relief forms and amounts could obviously impact the micropitting resistance. Besides, the increase of the profile modification amount would result in a lower profile form deviation, except some cases of very large modification amounts. Li [40] studied the effect of the angular misalignment on the micropitting resistance of a spur gear pair made of AISI 8620 steel. Numerical results showed that an optimized crown magnitude, which was dependent on both the input torque and the misalignment amount of the spur gear pair, should be determined for increasing the micropitting life. Weber et al. [41] from the FZG investigated the influence of the increased pressure angle on the flank surface capacity of gears by comparing the reference pressure angle (20°) and a modified tooth pressure angle (28°). Tests and calculations both indicated that when applying the same nominal contact stresses at the pitch point, the pressure angle modification might not have an evident impact on the micro-pitting load capacity. However, more experiments need to be designed on groups of different gear pressure angles to obtain higher relatively reliable results.

It is deserved to be mentioned that the effects of gear modification techniques on the gear contact characteristics have also been investigated by the author's team [42]. Liu et al. [43] proposed a mathematical concave modification model for the beveloid gear tooth surfaces with crossed axes. The loaded tooth contact analysis (LTCA) and the theoretical tooth contact analysis (TCA) were performed to the analytical contact patterns. In this work, four cases were presented and compared with each other: Case 1) Pinion with profile crowning modification and gear with profile concave modification; Case 2) Pinion without modification and gear with lead concave modification; Case 3) Both pinion and gear with lead concave modifications; Case 4) Pinion with combined concave modification and gear with combined crowning modification. The modification methods of all these four cases could enlarge the contact area, decreased the maximum value of the contact stress and increased the minimum film thickness compared with the case without any modification. That indicated the gear surface capacity such as micropitting resistance could be amplified by applying appropriate tooth modifications. The minimum film thickness, the contact pattern of TCA with and without modification, the contact pattern of LTCA with and without the modification of Case 4 (Pinion with combined concave modification and gear with combined crowning modification) are depicted in Figures 4 and 5, respectively.

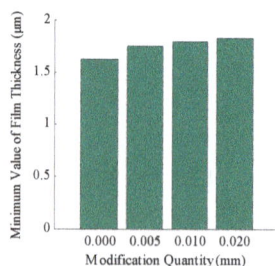

Figure 4. The minimum film thickness with different modification quantities of Case 4 (Reprinted with permission from [43]. Copyright 2018 SAGE).

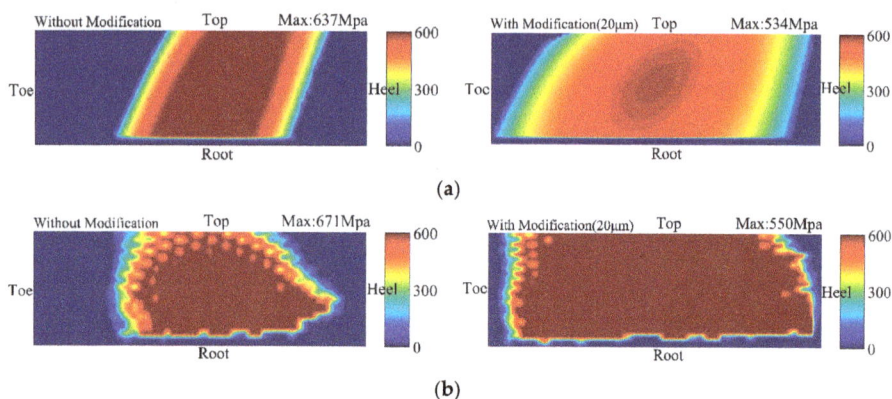

(a)

(b)

Figure 5. (a) The contact pattern of TCA with and without modification of Case 4; (b) The contact pattern of LTCA with and without modification of Case 4 (Reprinted with permission from [43]. Copyright 2018 SAGE).

2.2. Gear Surface Microscopic Topographies and Surface Treatments

Micropitting phenomena are closely associated with surface topography effects in high-duty hardened gears finished by grinding [44]. Berthe et al. [16] pointed out that surface roughness could be the driving force of the micropitting phenomenon, which is widely accepted. Therefore, various investigations have been conducted by many researchers to explore the effect of gear surface topography on micropitting [45,46]. Although the effect of surface topography on contact fatigue is often discussed under the lubrication state, the effects of lubrication properties (including the lubricant types, the additives, contaminates, etc.) are still expounded in the following section for a clearer explanation.

Winkelmann [6] from the NREL reported on the correlation between the surface roughness and micropitting, suggesting that the micropitting could be formed along the surface roughness asperities. Evans et al. [47] from Cardiff University, Cardiff, UK, predicted the micropitting damage through a micro-EHL model considering the measured surface roughness. The micropitting tests were carried out on helical gears for validation. In their work, the micropitting damages of generation-grinding and form-grinding gears were also compared, emphasizing the importance of 3D "waviness" on gear micropitting behaviors. Li et al. [48] proposed a primary numerical model considering lubrication and surface topography to predict the gear micropitting life. A year later, they improved the former model [49] and simulated the micropits distributions on the tooth flank. In addition, experiments were planned for the verification of the proposed model. A micropitting severity index (MSI) was defined as the cumulative probability in order to quantify the micropitting damage level. Whereas the effects of wear, case-hardening, residual stress gradient and the plastic deformation were still not considered in their work, which means a relative more comprehensive numerical model of micropitting is necessary to be further developed. Furthermore, Li [50] himself developed another numerical model a year later, which focused on the effect of surface roughness lay directionality on the micropitting of lubricated point contacts. The most severe micropitting damage was observed on the gear surface whose roughness lay was parallel to the rolling direction, while the roughness lay of its counterpart was vertical to the rolling direction. This conclusion has a considerable reference value on gear design and manufacturing processes, especially grinding and superfinishing. AL-Mayali et al. [51] investigated theoretically and experimentally the micropitting initiation based on the real surface topography under the micro-EHL regime. Both the numerical and experimental results showed that the micropits formed during the running-in stage (within 10^5 cycles) since the initial surface roughness asperities were significant. Mallipeddi et al. [52] compared the micropitting resistance of gears processed by grinding, honing and superfinishing. They concluded that the surface asperities could have a tremendous influence on the surface-initiated fatigue such as micropitting. The micropitting initiation could be significantly affected by the running-in load level, but only for rough surfaces.

Furthermore, the authors have numerically studied the effects of gear surface roughness on the RCF issues under the EHL condition considering the measured hardness and residual stress gradients, following the flow diagram shown in Figure 6.

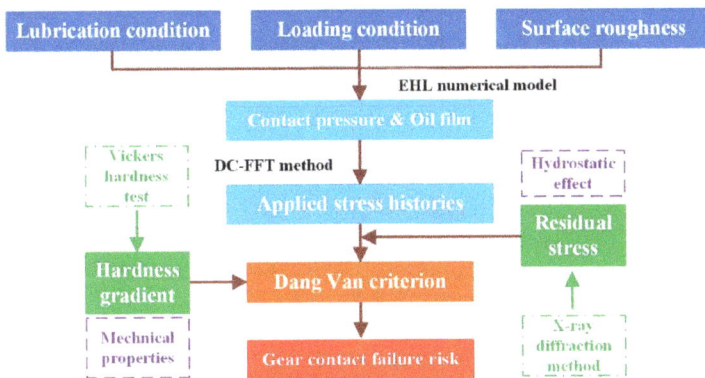

Figure 6. The technical route of contact failure risk evaluation with EHL considering different surface roughness states.

The carburized gear pair sample stems from the intermediate parallel stage of a 2 MW wind turbine gearbox, which is made of 18CrNiMo7-6. The main gear parameters are listed in Table 1. The carburizing, tempering, finishing and grinding processes were all adopted to obtain better material properties and higher failure resistance of this large-scale, heavy-load gear pair. The initial gear surface roughness, processed by the generating grinding method, was measured through a high-resolution optical device (Alicona G4) along the tooth flank profile. The initial roughness *RMS* value is around 0.25 μm. Figure 7 shows the optical measurement device and the initial surface roughness profile along the rolling direction.

Table 1. The gear parameters from the authors' studies.

Parameters	Values	Parameters	Values
Teeth Number	$Z_1 = 121$, $Z_2 = 24$	Pressure Angle	$\alpha_0 = 20°$
Normal Module	$m_0 = 0.011$ m	Helix Angle	$\beta_0 = 12°$
Shifting Coefficients	$x_1 = 0.034$, $x_2 = 0.4$	Contact Tooth Width	$B = 0.295$ m
Poisson's Ratio of Materials	$v_{1,2} = 0.3$	Young's Modulus	$E_{1,2} = 2.10 \times 10^{11}$ Pa
Transverse Pressure Angle	$\alpha_t = 20.41°$	Tip Clearance Coefficient	$c^* = 0.4$
Reference Input Torque of the Gear Pair	$T_{ref} = 241000$ N m	Reference Input Speed of the Gear Pair	$N_{ref} = 64.8$ rpm

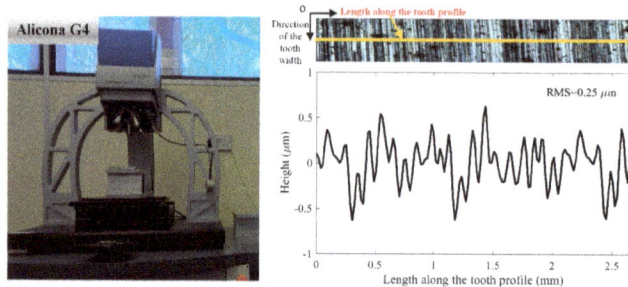

Figure 7. The optical profiler and the initial surface roughness profile.

The EHL theory was applied to predict the contact pressure distribution, addressing the effects of the surface roughness. The stress histories of interesting material points during complete contact loading cycles were calculated by exploring the discrete-concrete, fast Fourier transformation (DC-FFT) method [53]. Detailed EHL modeling and stresses calculation strategies can be referred to the authors' former work [54–56]. Subsequently, the Dang Van multiaxial fatigue criterion [57] shown below was applied to evaluate the multiaxial stress state during the contact, and the residual stress measured by the X-ray diffraction method could also be considered in the calculation.

$$\tau_{DangVan}(t) = \tau_{max}(t) + \alpha_D \cdot [\sigma_{H, load}(t) + \sigma_{H, Residual}(t)] \tag{3}$$

where $\tau_{Dang\ Van}$ is the Dang Van equivalent stress which can be deduced from the multiaxial stress history during the loading process, τ_{max} is the maximum shear stress, $\sigma_{H,load}$ means the hydrostatic stress caused by applied load, $\sigma_{H,Residual}$ represents the residual stress term. α_D is the material parameter which can be calculated through the hardness data measured by the Vickers hardness test [58]. The measured hardness and the residual stress gradients of the mentioned gear sample can be seen in Figure 8. It is deserved to be mentioned that the maximum amplitude of the induced compressive residual stress may not appear at surface [59,60], thus the effect of residual stress on micropitting was not addressed.

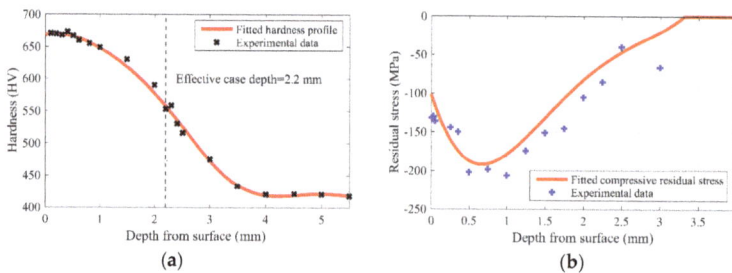

Figure 8. The fitted (**a**) hardness and (**b**) residual stress profiles based on the experimental method.

Based on the Dang Van criterion, a concept called the index of contact fatigue failure risk could be calculated as follows, in order to represent the failure possibility during the loading cycles:

$$R(z) = \frac{\tau_{\text{DangVan}}(z)}{\tau_{-1}(z)} \qquad (4)$$

where R is the risk of gear contact fatigue failure, τ_{-1} is the fully reversed torsion fatigue limit [61] which can be deduced from the hardness gradient, as expressed in Equation (5) [62]; and certainly, Z represents the depth.

$$\begin{cases} \tau_{-1}(z) = 0.773\text{HV}(z) - \text{HV}(z)^2/3170 \ (\text{MPa}) \\ \sigma_{-1}(z) = \tau_{-1}(z)/0.577 \ (\text{MPa}) \end{cases} \qquad (5)$$

As for the results, Figure 9 illustrates the risk of the contact fatigue along the depth at the position where the maximum τ_{max} is at peak value during the complete contact loading cycle within the calculation domain at the pitch point with different asperity conditions [63]. It reflects the fact that the increase of the surface roughness *RMS* makes the maximum index of the failure risk significantly rise, and its occurrence position gradually becomes shallower, reaching a depth of about 0.05 mm when the *RMS* rises to 0.5 μm, where the micropitting is apparently more likely to occur.

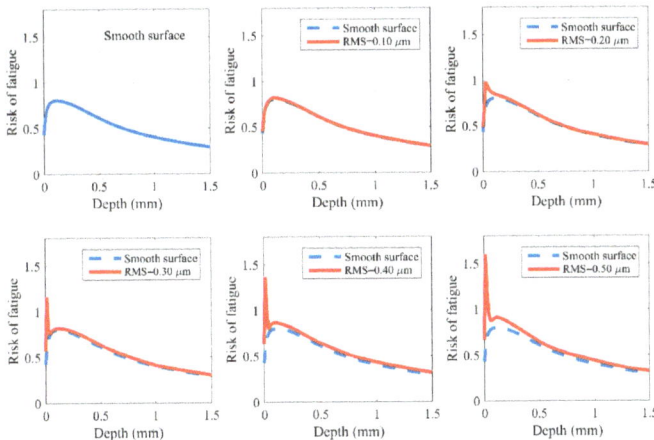

Figure 9. The index of contact fatigue failure risk along the depth with different *RMS* values [63].

Surface treatments such as grinding and superfinishing were proven to be effective on modifying the surface topographies positively, subsequently reducing the surface-initiated failure risk. As reported in Reference [6], micropitting could even be eliminated by superfinishing. Furthermore, superfinishing can also be applied to improve the lubricant life and cleanliness, which generates a higher component fatigue life.

Karpuschewski et al. [64] concluded two main reasons for gear finishing, (1) surface quality improvement and form error reduction, and (2) flank modification and surface integrity improvement. In order to quantify the effect of surface roughness on gear surface fatigue life, Krantz et al. [65] applied a near-mirror finish on AISI 9310 steel gears, the test data was also compared with the National Aeronautics and Space Administration (NASA) Glenn gear fatigue database. The results indicated that the superfinishing could extraordinarily reduce the surface roughness influence by making the roughness value R_a along the involute direction lower to around 0.1 μm, and subsequently leading to a higher micropitting resistance. Besides, the fatigue lives of gears with superfinishing were about 4 times greater than that of gears with ground teeth. Winkelmann et al. [66] mentioned that the

Isotropic Superfinish (ISF) gears during the fatigue tests never showed any sign of micropitting phenomena (micropitting coverage was 0%, profile form deviation was 0 μm) nor reached any of the specified failure criteria compared to the baseline gears with a lower micropitting resistance. From the gear micropitting experiments carried out by Rabaso et al. [67], a conclusion could be made that the ground gear surfaces obviously showed better micropitting phenomena than the turned gear surfaces. Shaikh et al. [68] applied electrochemical honing on bevel gears and found that within an optimized finishing time of 2 min, an obvious improvement in the surface finish and form accuracy were observed. Meanwhile, the gear surface integrity was improved and the surface failure resistance was enhanced. Ronkainen et al. [69] studied the influence of a surface finish under the high-load gear contacts. The results revealed that micropitting could be prohibited through polishing the gear surfaces by reducing approximately 27% of the friction coefficient.

Besides grinding and superfinishing, the effects of other surface treatments such as shot peening [70] and coating [69,71] are also widely studied, and strides have been made constantly on the understanding of these technologies. The diversity of investigations which focus on the effects of this kind of surface strengthening treatments on the gear micropitting behavior [72] are outlined. Terrin and Meneghetti [73] studied the contact fatigue on both shot-peened and un-peened 17NiCrMo6-4 case-hardening gear steel specimens through a two-disc test rig. No obvious improvement of micropitting resistance was found between the shot-peened and un-peened specimens. However, a contrary conclusion was presented by Pariente and Guagliano [74]. They analyzed the contact fatigue damage of shot peened gears through the X-ray diffraction method and claimed that the shot peening was effective to suppress the micropitting initiation. In addition, a similar conclusion could be drawn from the work of Rabaso et al. [67] mentioned above, in which it was stated that the shot peening was beneficial to the rolling contact fatigue (RCF) resistance. Koenig et al. [75] from the FZG, Germany, summarized the effects of shot peening on the tooth flank load carrying capacity reported in the FVA research project 521 I. Based upon the test results, they developed an extensive calculation method on the surface-initiated damage resistance, and claimed that both the shot peening and superfinishing could allow a positive effect on the gear surface properties, leading to a further increase of the gear tooth surface load capacity. According to the experiments conducted by Qin et al. [76], the ultrasonic nanocrystal surface modification (UNSM) has been proved to be a promising technique for the enhancement of micropitting, wear and pitting resistances of mechanical components such as gears and bearings, making the components surface roughness value R_a less than 0.02 μm.

With respect to the gear coating techniques, Krantz et al. [77] carried out experiments to compare the surface fatigue lives of case-carburized AISI 9310 steel spur gears with and without metal-containing, carbon-based coating. Almost all test coated gears survived from the fatigue failure after 275 million cycles and the coating was proved to be significantly effective to improve the tooth surface capacity. Vetter et al. [78] reviewed the improvements in surface treatments for automotive applications. A conclusion could be made that gear coated with a–C:H:Me (W–C:H) presented an obviously better micropitting resistance and higher load capacity. Martins et al. [79] discussed the effect of multilayer composite surface coatings (molybdenum disulphide/titanium (MoS2/Ti) and carbon/chromium (C/Cr)) on the gear capacity and friction coefficient. Experimental results showed that the load capacity of the coated gears had been highly improved. Bayón et al. [80] studied the gear tribological performance of the newly developed multilayer physical vapor deposition (PVD) coatings, namely Cr/CrN and CrN/ZrCN. They found that the Cr/CrN coating dramatically decreased the micropitting and scuffing damages. Moreover, the CrN/ZrCN coating could improve the wear resistance, especially under extreme pressure conditions. Moorthy and Shaw [81] from the Newcastle University, UK, tried to evaluate the contact fatigue performances of helical gears coated with Balinit C1000, Balinit C*, CrN + IFLM, C6 + IFLM and Nb–S coatings. The results suggested that the Nb–S coated gears have the best overall contact fatigue performance, followed by Balinit C coated gears with minimum micropitting damage. The same research team also compared the micropitting resistance of as-ground gears and gears coated with BALINIT® C and Nb–S coatings [82]. They concluded that both BALINIT® C and Nb–S coated gears possessed better micropitting behaviors by removing localized stress concentration at micro-valleys which presented on as-ground gears. Singh et al. [83] from the Argonne National Laboratory, US, developed a soft H-DLC coating (highly elastic and has a hardness value comparable to that of gear steel substrate) for sliding/rolling gear contact applications, which was proved to be beneficial to components that suffer from micropitting. During the tests, no failure was observed with the coated specimens even after 100 million cycles. As a comparison, the uncoated gear failed after only 32 million cycles. Benedetti et al. [84] investigated the effects of WC/C, WC/C–CrN, and DLC based coatings on lubricated contact behavior. The multi-layer WC/C and WC/CrN based coatings, together with the DLC coating showed a good contact fatigue performance, but these coatings were still unable to prevent the micro-pitting occurrence at the end of the tests.

2.3. Working Conditions

The influences of working conditions on micropitting have been widely investigated, and relative research could be found elsewhere [1,20,26]. This section reviews the studies which concern the effects of the normal load condition, the sliding/rolling condition and the lubrication state on micropitting behavior. Moreover, the corresponding recent results of the authors are presented.

2.3.1. The Normal Load Condition

The crucial influence of load condition on the micropitting has been addressed by many studies [25,85]. Fajdiga et al. [86] proposed a numerical model to simulate the micropitting growth on the tooth flank under boundary lubrication together with rolling and sliding conditions. This model was basically subjected to the normal contact pressure and the frictional forces, and the calculation results corresponded well with the experimental data. Olia and Bull [1] studied the diversity of influence factors on micropitting by experiments, including the normal load effect. They reported that micropitting initiation was mostly controlled by contact pressure and the micropitting propagation process was also mainly affected by the resultant pressure distribution. However, different conclusions of the loading influence on micropitting can be summarized from the diversity of studies. Webster and Norbart [87] reported that increasing applied load would accordingly increase the gear micropitting rate. While Rabaso et al. [67] found that the contact pressure would not impact the material's resistance to RCF in the range considered (1.5–2.5 GPa). Furthermore, Moallem et al. [88] estimated the micropitting life through the Zaretsky equation and applied the load-sharing concept, and subsequently comparing with the experimental results from the previous literature articles. The results showed that the applied load increased by about 4 times could decrease almost 50% of the micropitting life. Mallipeddi et al. [89] investigated the effect of running-in load state on surface characteristics of ground gears on an FZG rig. They summarized that a high running-in load would increase not only the operating efficiency but also the micropitting resistance. However, no obvious connection between the micropitting and the phase transformation within the gear steel was observed, which was inconsistent with the results obtained in the aforementioned studies. An et al. [90] demonstrated the micropitting occurrence through the Dark and Bright Ratio (DBR) method under different Hertzian contact pressures and developed a novel engineering regression equation to predict the RCF life.

Based upon the same gear sample and influence factors in the numerical model mentioned above, the loading effect was discussed by the authors. Several input working conditions of this parallel stage were chosen from the measured load spectrum, including the low input loading case (input torque = 102,000 N m), the medium loading case (reference input torque = 241,000 N m) and the high loading case (input torque = 363,000 N m), all the input speeds of these three cases were set to be the reference input speed, which is 64.8 rpm as listed in Table 1 for clear comparison. The contact fatigue lives were estimated by applying the Brown–Miller multiaxial criterion [91] modified with the mean stress as shown below; this multiaxial criterion is based on the critical plane criterion:

$$\frac{\Delta \gamma_{max}}{2} + \frac{\Delta \varepsilon_n}{2} = C_1 \frac{(\sigma'_f - \sigma_m)}{E} (2N_f)^b + C_2 \varepsilon'_f (2N_f)^c \tag{6}$$

where $\Delta\gamma_{max}/2$ stands for the maximum amplitude of the shear strain, $\Delta\varepsilon_n/2$ refers to the amplitude of the normal strain on the critical plane where the maximum amplitude of the shear strain appears, σ_m represents the mean of the normal stress on the critical plane; b, c are the fatigue strength index and the ductility index, respectively; C_1, C_2 are material constants, N_f is the number of cycles of fatigue; σ'_f, ε'_f are the axial fatigue strength coefficient and the axial fatigue ductility coefficient, respectively. The schematic diagram of the calculation domain and the critical plane for the Morrow–Brown–Miller multiaxial criterion are shown in Figure 10. Where the calculation domain can be divided into many equally spaced grids for the sake of both the calculation efficiency and accuracy. Parameter θ is the angle between the direction of the normal stress σ_n and the positive direction of the x-coordinate. With the view to determining the critical plane where the maximum amplitude of the shear strain takes place, all the candidate planes had to be scanned with a 5° step from 0° to 180° at each interested material point within the calculation domain.

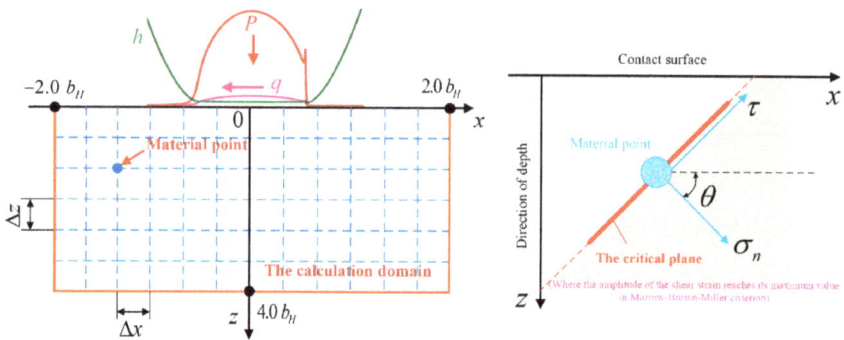

Figure 10. The schematic diagram of the calculation domain and the critical plane of RCF.

Figure 11 shows the contact fatigue lives distributions along the depth calculated based on different input torques with the initial surface roughness. As can be seen clearly, the minimum surface fatigue life decreases from 1.42×10^6 to 2.29×10^5 significantly as the input torque increases. Moreover, the minimum fatigue lives all appear at the positions very near to the gear surface in these three cases, indicating that the micropitting risk may be predominant at this moment. The result implies that larger loading may lead to faster micropitting damage, namely, a shorter micropitting fatigue life, which agrees well with the studies in reference [88].

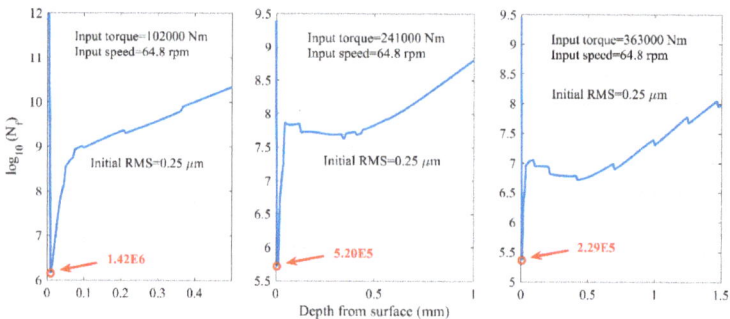

Figure 11. The contact fatigue lives under different input loadings of the first loading cycle with initial surface roughness.

2.3.2. The Sliding/Rolling Condition

The sliding/rolling condition is a general factor of the gear micropitting phenomena, which is being widely explored in many studies [82,92]. According to the work implemented by Olia and Bull [1], speed and slide-roll ratio have the biggest effects on micropitting propagation. Furthermore, the micropitting crack angle and propagation direction could be directly influenced by the sliding/rolling condition [85,93]. Figure 12 schematically presents the correlation between the micropitting crack growth direction and the sliding/rolling condition on the gear tooth flank of the driving and the driven gears.

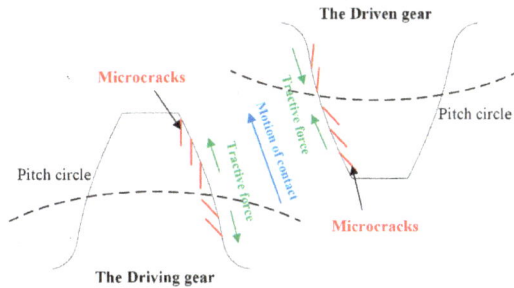

Figure 12. The correlation between the micropitting crack propagation direction and the gear sliding/rolling condition (Reprinted with permission from [85,93]. Copyright 2005 SAGE & 2012 Gear Technology).

Speed is one of the main influencing parameters that can be controlled and studied during the gear experiments [88]. Based on Seireg's research [94], a higher gear speed may result in smaller size and shallower depth of the micropits. Interestingly, this paper also considers the effects of thermal shock, thermal stress cycling, coatings, surface roughness and different surface treatments on the pitting, micropitting and wear behaviors.

The value of the slide-roll ratio (SRR) illustrates the proportion of the sliding effect during the gear contact, which is very crucial to the evaluations of gear surface capacities, such as micropitting, wear and scuffing resistance. A few decades ago, Zhou et al. [95] from the Northwestern University, US, proposed a numerical model for predicting the onset of gear micropits for a given lubrication state under rolling and sliding contacts condition. The calculation results were also verified by experimental data obtained through a two-disk rig. Errichello [96] illustrated that the micropitting crack growth could be affected by the rolling and sliding condition. The micropitting cracks usually grow opposite the sliding direction at the gear tooth surface. From the explanation of Morales-Espejel and Gabelli [94] from the Svenska Kullager-Fabriken (SKF) Engineering and Research Centre, the Netherlands, the surface spalling direction may depend on the friction direction of the sliding-rolling condition, instead of random directions under pure rolling condition. Sanekata et al. [97] reproduced the micropitting caused by RCF on the carburized SCM420 steel under applied SRR of 0%, −20% and −40%. A large number of micropits appears after the RCF tests. The results showed that the angle of microcracks between the initiation and the rolling direction increases as the SRR, as well as the shear strain, increases. Another reference from Morales-Espejel et al. [46] selected three SRR values, 0.05, 0.15, 0.30, in order to emphasize the important role of SRR on the micropitting phenomena. They concluded that the micropitting occurrence position could be affected by the combined effects of SRR, contact pressure and wear phenomenon. The micropits might appear inside the single pair contact region due to the combined effects, rather than the positions where the maximum contact pressure exists, as generally considered. Cen et al. [98] from the University of Leeds, UK, tried to investigate the effect of SRR on the micropitting damage. The test gears were lubricated with Zinc Dialkyl Dithiophosphate

(ZDDP) containing lubricants. The results showed that increasing SRR can reduce the micropitting area on the worn surface obviously.

The influence of sliding/rolling condition on the surface-initiated failures can also be reflected in the authors' recent work. Based on the aforementioned numerical model, the contact analysis and the wear investigation could be launched primarily. The Archard's wear model is a well-accepted wear prediction method [99–101] which is quite appropriate to reveal the influences of contact parameters on the wear process. The general Archard's wear model of each local material point on one of the interacting surfaces can be formulated as [102]:

$$\frac{\mathrm{d}h_{\mathrm{w}}}{\mathrm{d}s} = kp_{\mathrm{m}} \tag{7}$$

where h_{w} stands for the wear depth, s denotes the sliding distance, p_{m} represents the mean value of the local contact pressure, and the wear coefficient k is chosen to be $1.9 \times 10^{-19} \mathrm{~m^2~N^{-1}}$ according to reference [103], where the similar gear steel material after case-hardening is used. Based on the same gear sample introduced in Section 2.2, Figure 13 displays the maximum Hertzian contact pressure p_{H} and the SRR distributions along the LOA [104].

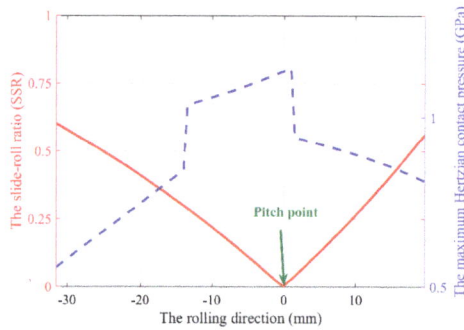

Figure 13. The maximum Hertzian contact pressure and the SRR along the LOA [104].

Moreover, Figure 14 presents the estimated wear depths along the line of action (LOA) during the first loading cycle of both the driving and the driven gears [104]. Compared with these two figures, a simple conclusion can be made that the micropitting, wear behaviors and the SRR are interrelated. Apparently, the contact pressure, the sliding/rolling condition and the wear coefficient are the main factors on the surface contact fatigue performance according to the Archard's wear equation. In this study, the largest wear depth appeared at the lowest point of single tooth contact (LPSTC) of the driven gear.

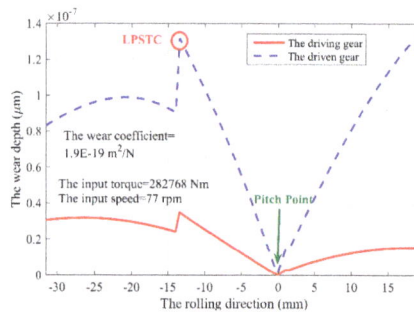

Figure 14. The wear depths of the driving and the driven gear along the LOA [104].

2.3.3. Lubrication Conditions

The earliest discussion of micropitting in the tribology field appeared in the study by Berthe, et al. [16], and the lubrication condition was generally the main concern of this kind of studies [105]. The micropitting phenomenon is generally analyzed under the mixed or boundary lubrication condition [106,107]. Moreover, the lubricants type [93], the additives [108], and other lubricant properties [109] can have noticeable influences on the contact fatigue behavior. In the following studies, the effect of lubricant types, additives, oil supply condition and the thermal effect on micropitting are outlined.

Different types of the lubricants used in the gear contact can result in a total discrepancy on micropitting damage [110]. The discussion of the effects of lubricant types on micropitting is mainly between the mineral oils and ester oils. According to Martins et al. [31], the direct comparison of biodegradable ester and mineral oils effects on the micropitting of austempered ductile iron gears was performed. The gear micropitting tests were conducted in an FZG test rig using these two types of industrial gear oils samples, and they found that compared to the gears lubricated with mineral oil, the gears lubricated with ester oil possessed a 20% lower cumulative mass loss. Additionally, the micropitting area of these two cases was similar. According to the gear micropitting performed by the same research team [111], still no obvious discrepancy on the micropitting phenomena was found between the carburized gears using the mineral and biodegradable ester gear oils. Cardoso et al. [112] compared the gear micropitting behavior of high-pressure nitriding steel gears which were lubricated with a standard mineral lubricant and two biodegradable esters. The gears with the ester lubricants presented better a micropitting resistance than the gears with the mineral oil. This study may support the applicability of biodegradable ester oils on gear lubrication, mainly when the high micropitting resistance is required.

The lubricants are generally composed of a majority of base oils plus a variety of additives to impart desirable characteristics and the composition of lubricants have been extensively investigated. The properties of the additives can directly influence the fatigue failure resistance of the gear surface. Lainé et al. [113] added a common friction modifier agent, molybdenum bis-diethylhexyl dithio-carbamate (MoDTC), to an oil containing an anti-wear additive (secondary ZDDP) and to a mineral base-stock in order to study the influence of the friction modifier additive on micropitting. The micropitting could be greatly reduced when added with ZDDP, compared to the samples added with MoDTC. Ochoa et al. [114] explored the effects of the different additive types in a polyalphaolefinic low-viscosity base oil, namely a polyalphaolefin PAO6, on the gear micropitting resistance. The results suggested that the extreme pressure (EP) or anti-wear (AW) additives might lead to a greater micropitting damage on the gear surfaces at the running-stage under severe working conditions, such as heavy loading and higher roughness. Soltanahmadi et al. [115] studied the effect of N-tallow-1, 3-diaminopropane (TDP) on micropitting in order to develop lubricants with no or minimal environmental impact. The results revealed that the TDP-containing lubricant was effective in mitigating the micropitting damage.

The contamination could be regarded as an unconventional kind of additive. With the view of estimating the effect of water contamination in gear lubricants on the micropitting and wear of carburized gears, an FZG gear test rig was employed in Reference [116], and the interaction lubricant-water (effect of water on the molecular structure of base oils and additives), chemical-material-technological (especially corrosive reactions) as well as tribological influence (effect of water droplets in the contact zone) were investigated. No obvious influence of the water contaminations on the tooth micropitting and wear resistance values was observed during all experiments. Moreover, it is worthy to be mentioned that plenty of patents invented by researchers have tried to determine the appropriate composition of industrial oil to control the gears surface micropitting damage [117,118].

The lubricated method or the oil supply method also plays an important role in lubrication condition. The gear micropitting resistance varies significantly with the oil supply condition.

Liu et al. [56] numerically investigated the contact performance of a spur gear pair under the starved lubrication condition. The minimum film thickness along LOA might appear at the area around the HPSTC point, compared to the engage-in point of full film cases, indicating that the starved lubrication state would influence the occurrence position of micropitting. Moss et al. [119] employed experiments to investigate the effect of lubrication methods on the spur gear contact fatigue performance. From the test results, they concluded that no tangible discrepancy of the micropitting under varying dip and jet-lubricated conditions was observed, but the spin power losses were significantly affected by the lubrication methods.

The change of temperature can directly lead to a remarkable variation of the lubrication performance because the viscosity and the chemical activity of the lubricants are quite sensitive to thermal fluctuation. Taking the thermal phenomenon into consideration of the gear design would give better insight into the mechanisms of contact fatigue failure modes [120]. Höhn and Michaelis [21] studied the effect of oil temperature on several contact fatigue failures including micropitting. For most test lubricants, the micropitting durability dropped with decreasing viscosity and film thickness, which corresponded to a higher oil temperature. The thermal effect on micropitting has not been individually or emphatically considered in the gear fields, especially in numerical modeling investigations. Most studies evaluated the thermal influence through controlling the test temperatures [1,101], or through the influence on the lubrication behaviors, just like the work of Höhn and Michaelis mentioned above. Hence, in view of the significant impact of the thermal effect on micropitting, further explorations need to be launched.

Experiments have also been implemented by the authors to investigate the effects of the working conditions, especially the surface rolling speed and the specific lubricant film thickness, on the micropitting phenomenon [121]. The experiments were conducted by a PCS Instruments micropitting rig (MPR) shown in Figure 15, which possesses a large range of SRR. The three discs (ring) and the roller are made of AISI 52100 with surface hardness values of 60 HRC and 63 HRC, respectively. The contact track width of the roller is 1 mm, which is also displayed in Figure 15 [121]. During the operation, the contact stresses of these cases were all set to be 1.9 GPa, while different surface rolling speed values u_r and the specific lubricant film thickness λ values were applied to study their influences on micropitting damage.

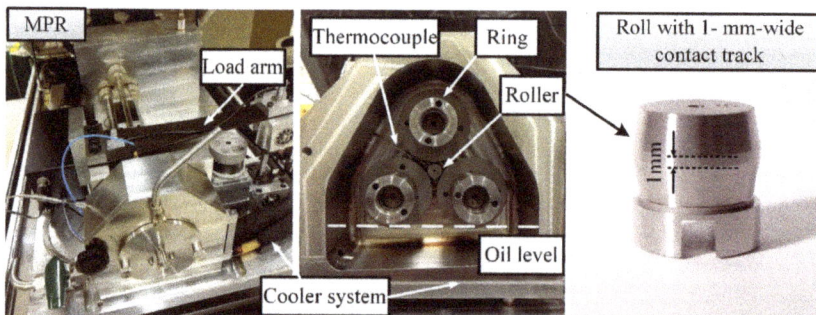

Figure 15. The PCS Instruments triple-contact micropitting rig and the test roller [121].

Figure 16 gives the results of the damaged surfaces of test 1-4, the relative test parameters are also listed [121]. The specific lubricant film thickness of these tests are 0.06, 0.11, 0.17, 0.25, respectively. Figure 16a,b represent the micropitting damage surfaces under the low-speed condition, $u_r = 1$ m s^{-1}, where no intensive micropitting was observed. As for the higher-speed condition, test 3 and 4 results with $u_r = 3.4$ m s^{-1} (shown in Figure 16c,d), show that the micropits area was much larger than that of the lower-speed condition. Moreover, the micropitting damage could even be observed on most of the surface in test 3. This phenomenon might indicate that the micropitting damage could commonly be observed at higher speeds.

While the influence of the specific lubricant film thickness λ is not consistent with the general understanding, theoretically, a larger specific lubricant film thickness λ means a better lubrication condition. However, under the lower-speed condition, test 2 with a higher λ (0.11) obviously showed more severe micropitting damage than test 1 (λ = 0.06) until 38 million cycles. This phenomenon could be explained by considering the wear process. For lower λ values (like test 1), the surface roughness could be worn significantly within the running-in stage, leading to decreasing the micropitting damage accumulation gradually. Hence, after 38 million cycles, the surface of test 1 might suffer less micropitting damage than test 2, with higher specific lubricant film thickness values because of the wear influence.

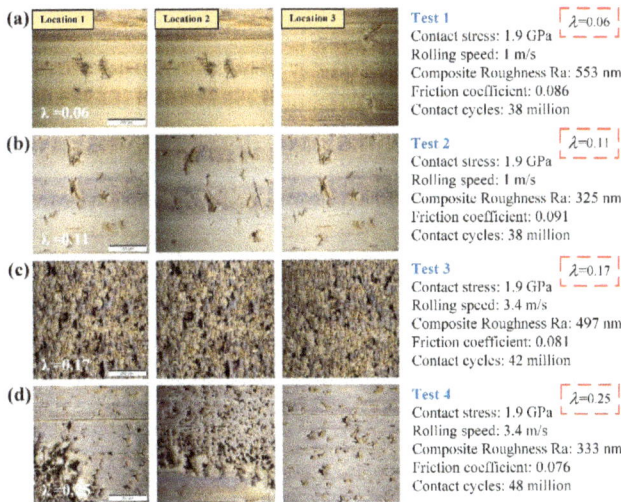

Figure 16. The roller surface and operating parameters of tests 1–4 (a–d) [121].

The above discussion demonstrates the particular importance of considering the wear effect when analyzing the micropitting mechanism instead of applying the contact fatigue model solely. A detailed introduction and explanation are carried out in the following section.

3. The Contact Fatigue Failure Competitive Mechanism Considering Wear Effect

When investigating the micropitting mechanism, the majority of the aforementioned works do not consider another important factor, namely the wear effect. During the cycling contact, the main loss of the material is caused by the wear process [2], changing the gear surface topography simultaneously. This may lead to the alternate occurrences among the several contact failure modes, which could be regarded as a failure competitive mechanism. Consequently, the effect of the wear process on the contact fatigue performances should also be taken into account in order to draw relatively accurate conclusions [46].

3.1. Review on the Studies Considering the Competitive Mechanism

The recent progress in the study of the competitive mechanism among contact fatigue failures and wear damage is mainly attributed to the works of the research teams of Morales-Espejel and Brandão.

In 2011, Morales-Espejel and Brizmer [122] established a model for the micropitting of rolling bearings under the sliding-rolling condition. In their work, a concept called competitive mechanism between the surface fatigue failures and mild wear was first raised. The effects of roughness, lubrication, sliding, etc. were considered and the wear process was simulated based on the Archard's model [123]. Morales-Espejel et al. [46] explained that the existence of wear could move the micropits inside the single pair contact region, rather than to the extremes positions when the wear was not taken into account.

Brandão et al. [2] developed a numerical method to simulate the concurrent gear micropitting and mild wear. However, they just considered the Archard's wear model and the surface fatigue models in succession only and summarized that combining the wear model and surface fatigue model was very important to develop a good micropitting model.

In addition, Vrcek et al. [107] assessed the micro-pitting and wear performances of engine oils under the boundary lubrication condition. Based on the test results, they claimed that micropitting and mild wear were competing phenomena and that higher micropitting damage would result in lower wear damage, which corresponds well with the previous research [108,124].

It is worth noting that some studies also mentioned the competitive mechanism between mild wear and the surface-initiated damage through indirect explanations, mainly stemming from investigations of micropitting phenomena with extreme pressure (EP) or anti-wear (AW) additives. For instance, Benyajati et al. [125] designed micropitting tests by applying a new miniature test-rig. They stated that certain anti-wear additives, like ZDDP, might be detrimental to micropitting resistance since they prevented wear (decline the smoothing rate of surface roughness) rather than by any direct influence on micropitting. As recorded in Reference [85], the mild wear would smoothen the surface roughness and reduce the micropitting risk simultaneously, suggesting that other failure modes like pitting might play a dominant role during meshing. However, when the anti-wear additive was applied, the initial surface roughness persisted, leading to severe micropitting. However, O'Connor [126] explained that EP or AW additives should not be simply defined as beneficial or detrimental on micropitting resistance since the chemical structure and the interaction effects of the additives could remarkably influence micropitting through a complicated way. Therefore, more detailed tribochemical studies of lubricants additives on the gear RCF and micropitting performance are needed.

3.2. Progress on Competitive Mechanism

In this regard, the authors also studied the gear contact failure competitive mechanism resulting from wear process [104]. Instead of revealing the competitive phenomenon between the contact failures and wear, the competitive mechanism between micropitting and pitting due to the wear effect was investigated. According to Figure 14, the wear depth accumulation could be primarily predicted on the LPSTC of the driven gear based on the initial surface roughness through the Archard's wear model [123].

Based on the aforementioned gear pair sample and numerical model in Section 2, combined with the Morrow–Brown–Miller multiaxial criterion and the Palmgren–Miner rule [127], the competitive mechanism between micropitting and pitting considering the wear process could be primarily investigated through the damage accumulation distributions. The Palmgren–Miner rule can be expressed as follows:

$$D = \sum_{i=1}^{m} D_i = \sum_{i=1}^{m} \frac{n_i}{N_i} \leq 1.0 \qquad (8)$$

where D is the total damage accumulation during the loading cycles, D_i means the damage suffered during a loading block, n_i denotes the actual number of cycles experienced under the current loading stage, N_i represents the fatigue life at the corresponding constant amplitude loading condition calculated by using the Morrow–Brown–Miller criterion. The total damage accumulation won't exceed the threshold value of 1.0 before the final failure occurs [61].

However, it is quite time-consuming when estimating the damage accumulation of each loading cycle because of the high-cycle fatigue characteristics of the gears. Hence a method called "jump-in-cycles" reported in Reference [128] was applied for both the calculation accuracy and efficiency. Based on this method, the gear surface roughness could be assumed to remain unchanged over a finite loading block ΔN. In the authors' work, the loading cycles number within a loading block was decided to be $\Delta N = 1 \times 10^6$.

Figure 17 shows the surface roughness profile of different damage stages [104]. As the wear process proceeds, the volumetric and the smooth wear effects on the surface roughness would occur simultaneously. Therefore, in order to guarantee the relative accuracy of the surface roughness *RMS* evaluation, the reference lines of these worn surface roughness profiles at different damage stages along the rolling direction are adjusted to be the *x*-coordinate in order to meet the *RMS* calculation requirement. During the wear process, the surface roughness decreases from 0.25 µm (initial stage) to about 0.07 µm (when the maximum damage accumulation reaches 1.0).

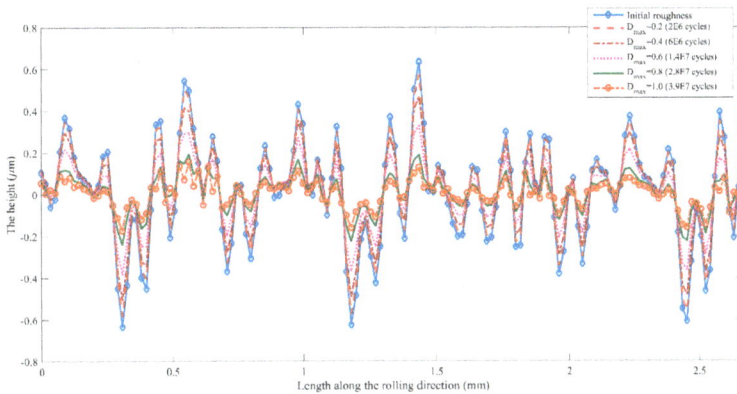

Figure 17. The surface roughness profile of different damage stage with the adjustment of the reference lines of the driven gear [104].

Figure 18 displays the damage accumulation within every certain damage stage ($\Delta D = 0.2$) [104]. The maximum damage accumulation appears at about 0.03 mm near the surface, where micropitting plays a dominant role compared with pitting failure. However, as the wear process proceeds, the depth of the maximum damage increases gradually. Consequently, the subsurface failure risk like pitting risk becomes predominant as can be seen when the total damage accumulation *D* reaches 1.0, the maximum damage accumulation occurs at a depth of about 0.28 mm. This indicates that the micropitting risk would gradually change to the pitting risk when the wear is considered.

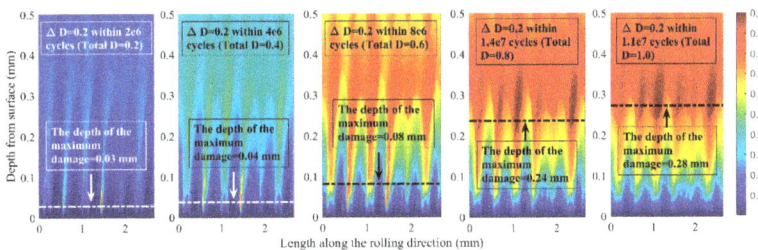

Figure 18. The damage accumulation within every individual certain damage stage ($\Delta D = 0.2$) [104].

Additionally, the effect of the initial surface roughness on the competitive mechanism has also been studied. Figure 19 displays the depth of the maximum damage accumulation (when $D = 0.2$) of different initial surface roughness values from 0.07 to 0.25 µm [104]. The depth of the maximum damage accumulation decreases from around 0.28 mm (initial roughness *RMS* = 0.07 µm) to around 0.027 mm (initial roughness *RMS* = 0.25 µm). The region around 0.1 mm could be set as the critical region to distinguish the occurrences of pitting and micropitting [9]. Hence, when the initial surface *RMS* is less than about 0.18 µm, the micropitting may completely disappear during the whole loading cycle, which agrees well with the results of the aforementioned studies.

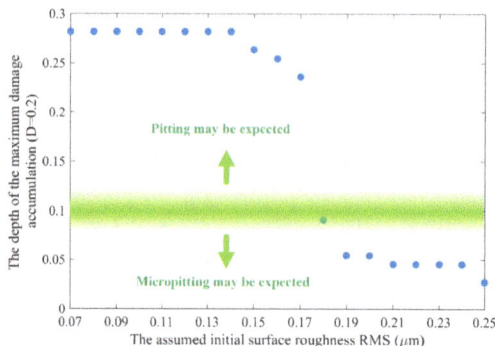

Figure 19. The depth of the maximum damage accumulation (when $D = 0.2$) of different initial surface roughness values [104].

Furthermore, the numerical results were also verified through the experimental results. Figure 20 displays the evolutions of both the measured roller surface roughness and the simulated octahedral shear stress from 1 to 40 million cycles [121]. The Hertzian center represents the contact center of the Hertzian contact width based on the Hertzian contact theory. During the first 20 million cycles, the wear effect on the surface roughness is significant. A similar trend of variation could also be observed on the octahedral shear stress distributions. The wear can obviously reduce the surface roughness asperities within the running-in stage, subsequently mitigating the micropitting damage to a certain extent. This would lead to a competitive phenomenon since the micropitting is suppressed by the wear. Meanwhile, other failure modes such as pitting or tooth flank fracture may become more dominant gradually.

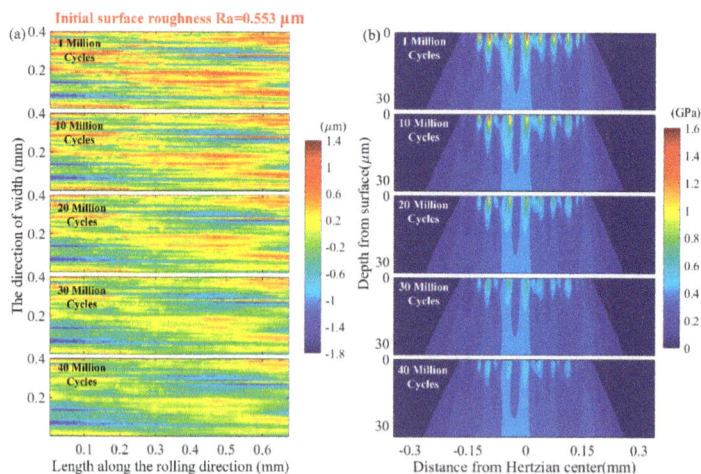

Figure 20. (**a**) The evolution of surface roughness and (**b**) the evolution of octahedral shear stress from 1 million to 40 million cycles [121].

4. Conclusions

In this work, plenty of relevant studies on the micropitting of steel gears, especially the competitive phenomena that occurs among several contact fatigue failure modes when considering gear tooth surface wear effect, are reviewed. Meanwhile, some recent research results about gear micropitting obtained by the authors are also illustrated for more comprehensive discussions. Several conclusions can be summarized as follows:

- The gear micropitting resistance can be influenced by both macroscopic and microscopic geometry factors. The appropriate gear tooth modification method is beneficial to the anti-micropitting capacity. The increase of the surface roughness *RMS* makes the maximum index of failure risk significantly increase and its occurrence position comes closer to the surface (about 0.05 mm when *RMS* = 0.5 μm), where the micropitting is more likely to occur. However, the superfinishing may completely eliminate micropitting due to mitigating the effect of surface roughness.
- The micropitting initiation is mostly controlled by contact pressure, whereas the propagation process can be significantly influenced by the sliding/rolling condition, and the micropitting damage may commonly be observed at higher speeds. With the increase of the specific lubricant film thickness from 0.06 to 0.17, the micropitting damage becomes more severe unless the R_a of the surface roughness is relatively low. The extreme pressure (EP) or anti-wear (AW) lubricant additives may be detrimental to the micropitting resistance because the surface roughness can be protected from wearing, leading to greater micropitting damage at the running-in stage.
- Merging the effects of the wear process and surface fatigue failure damages of the gear tooth surface is essential for the successful gear micropitting modelling. During the wear process, the position of the maximum damage accumulation moves gradually from near-surface to subsurface, indicating that the pitting failure is becoming more dominant than micropitting. The occurrence probability of micropitting can be significantly reduced due to the existence of the wear, resulting in the competitive mechanism between micropitting and pitting during the loading cycles.

Although impressive progress has been achieved on gear micropitting, and numerous factors have been considered in gear contact fatigue analysis [129–131], the micropitting mechanism still has not been fully revealed. The combined consideration of materials, structures, working condition, chemistry factors, etc., is much more necessary to numerical and experimental gear micropitting investigations for both the academic study and the industry.

Funding: This research was funded by National Natural Science Foundation of China (Nos. 51775060, 51805049, U1864210).

Conflicts of Interest: The authors declare no conflict of interest. The funders had no role in the design of the study; in the collection, analyses, or interpretation of data; in the writing of the manuscript, or in the decision to publish the results.

Nomenclutare

AW	Anti-wear
b, c	The fatigue strength index and the ductility index, respectively
C_1, C_2	The constants used in Morrow–Brown–Miller criterion
B	Contact tooth width, m
c^*	The tip clearance coefficient
D	The damage accumulation
E_1, E_2	Young's modulus of the driving and the driven gears, respectively, Pa
EHL	Elastohydrodynamic lubrication
EP	Extreme pressure
h_0	The lubricant film thickness with the assumption that the contact surface is smooth, μm

h_w	The wear depth, m
k	The wear coefficient, $\mathrm{m_2\,N^{-1}}$
LOA	Line of action
LPSTC	The lowest point of single tooth contact
m_0	The normal module of the gear pair, m
MPR	Micropitting rig
MSI	Micropitting severity index
N_ref	The reference input speed of the driving gear, rpm
N_f	The number of cycles of fatigue
p_H	The maximum Hertzian contact pressure, Pa
p_m	The mean value of the local contact pressure, Pa
RA	The retained austenite
RCF	Rolling contact fatigue
RMS	The root-mean-square, µm
R_a	The effective arithmetic mean roughness value, µm
R_q1, R_q2	The *RMS* of contact surfaces of the driving and the driven gears, respectively, µm
R_q	The combined *RMS* value, µm
s	The relative sliding distance, m
S_λ	The safety factor representing the micropitting load capacity
$S_{\lambda,\,\mathrm{min}}$	The minimum required safety factor
SRR	The slide-roll ratio
T_ref	The reference input torque of the driving gear, Nm
x_1, x_2	The shifting coefficients of the driving and the driven gears, respectively
Z_1, Z_2	The teeth number of the driving and the driven gears, respectively
ZDDP	The Zinc Dialkyl Dithiophosphate
α_0	The pressure angle of the gear, °
α_D	The material parameter used in Dang Van criteria
α_t	The transverse pressure angle of the gear, °
β_0	The helix angle of the gear, °
σ_f'	The axial fatigue strength coefficient, MPa
σ_m	The mean of the normal stress at the critical plane, MPa
$\Delta\gamma_\mathrm{max}/2$	The maximum amplitude of the shear strain
τ_DangVan	The Dang Van equivalent stress, MPa
τ_max	The maximum shear stress, MPa
ε_f'	The axial fatigue ductility coefficient
$\Delta\varepsilon_\mathrm{n}/2$	The amplitude of normal tensile strain at the critical plane
$\upsilon_{1,\,2}$	The Poisson's ratio of materials
λ	The specific lubricant film thickness
λ_min	The minimum specific lubricant film thickness in the contact area
λ_p	The permissible specific lubricant film thickness

References

1. Oila, A.; Bull, J. Assessment of the factors influencing micropitting in rolling/sliding contacts. *Wear* **2005**, *258*, 1510–1524. [CrossRef]
2. Brandão, J.; Martins, R.; Seabra, J.; Castro, M. An approach to the simulation of concurrent gear micropitting and mild wear. *Wear* **2015**, *324–325*, 64–73. [CrossRef]
3. Terrin, A.; Dengo, C.; Meneghetti, G. Experimental analysis of contact fatigue damage in case hardened gears for off-highway axles. *Eng. Fail. Anal.* **2017**, *76*, 10–26. [CrossRef]
4. Höhn, B.-R.; Oster, P.; Emmert, S. Micropitting in case-carburized gears-FZG micro-pitting test. *VDI Berichte* **1996**, *1230*, 331–344.
5. Sheng, S. *Wind Turbine Micropitting Workshop: A Recap*; National Renewable Energy Laboratory: Golden, CO, USA, 2010.

6. Winkelmann, L. *Surface Roughness and Micropitting*; National Renewable Energy Laboratory, Wind Turbine Tribology Seminar: Golden, CO, USA, 2011.

7. Martins, R.; Locatelli, C.; Seabra, J. Evolution of tooth flank roughness during gear micropitting tests. *Ind. Lubr. Tribol.* **2011**, *63*, 34–45. [CrossRef]

8. Brandão, J.; Seabra, J.; Castro, J. Surface initiated tooth flank damage: Part I: Numerical model. *Wear* **2010**, *268*, 1–12. [CrossRef]

9. Brandão, J.; Seabra, J.; Castro, J. Surface initiated tooth flank damage. Part II: Prediction of micropitting initiation and mass loss. *Wear* **2010**, *268*, 13–22. [CrossRef]

10. Whitby, R. Micropitting: An engineering or chemistry problem? *Tribol. Lubr. Technol.* **2004**, *60*, 56.

11. Winter, H.; Oster, P. *Influence of the Lubricant on Pitting and Micro Pitting (Grey Staining, Frosted Areas) Resistance of Case Carburized Gears: Test Procedures*; American Gear Manufacturers Association: Alexandria, VI, USA, 1987.

12. Brechot, P.; Cardis, A.; Murphy, W.; Theissen, J. Micropitting resistant industrial gear oils with balanced performance. *Ind. Lubr. Tribol.* **2000**, *52*, 125–136. [CrossRef]

13. Sun, Y.; Bailey, R. Effect of sliding conditions on micropitting behaviour of AISI 304 stainless steel in chloride containing solution. *Corros. Sci.* **2018**, *139*, 197–205. [CrossRef]

14. Way, S. Pitting due to rolling contact. *J. Appl. Mech.* **1935**, *2*, A49–A58.

15. Dawson, P. Effect of metallic contact on the pitting of lubricated rolling surfaces. *J. Mech. Eng. Sci.* **1962**, *4*, 16–21. [CrossRef]

16. Berthe, D.; Flamand, L.; Foucher, D.; Godet, M. Micropitting in Hertzian contacts. *J. Lubr. Technol.* **1980**, *102*, 478–489. [CrossRef]

17. Snidle, R.; Evans, H. Elastohydrodynamics of gears. *Tribol. Ser.* **1997**, 271–280. [CrossRef]

18. Olver, A. Gear lubrication-a review. *Proc. Inst. Mech. Eng. Part J J. Eng. Tribol.* **2002**, *216*, 255–267. [CrossRef]

19. Tallian, T. On competing failure modes in rolling contact. *ASLE Trans.* **1967**, *10*, 418–439. [CrossRef]

20. *ISO/TR 15144-1:2010 Calculation of Micropitting Load Capacity of Cylindrical Spur and Helical Gears-Part 1: Introduction and Basic Principles*; ISO: Geneva, Switzerland, 2010.

21. Höhn, B.-R.; Michaelis, K. Influence of oil temperature on gear failures. *Tribol. Int.* **2004**, *37*, 103–109. [CrossRef]

22. Höhn, B.-R.; Oster, P.; Radev, T.; Steinberger, G.; Tobie, T. Improvement of standardized test methods for evaluating the lubricant influence on micropitting and pitting resistance of case carburized gears. In Proceedings of the AGMA Fall Technical Meeting, Orlando, FL, USA, 22–24 October 2006.

23. Houser, D.; Shon, S. An Experimental Evaluation of the Procedures of the ISO/TR 15144 Technical Report for the Prediction of Micropitting. In Proceedings of the AGMA Fall Technical Meeting, Detroit, MI, USA, 2–4 October 2016.

24. Long, H.; Al-Tubi, I.; Martineze, M. Analytical and Experimental Study of Gear Surface Micropitting due to Variable Loading. *Appl. Mech. Mater.* **2015**, *750*, 96–103. [CrossRef]

25. Al-Tubi, I.; Long, H.; Zhang, J.; Shaw, B. Experimental and analytical study of gear micropitting initiation and propagation under varying loading conditions. *Wear* **2015**, *328*, 8–16. [CrossRef]

26. *ISO/TR 15144-1:2014 Calculation of Micropitting Load Capacity of Cylindrical Spur and Helical Gears-Introduction and Basic Principles*; ISO: Geneva, Switzerland, 2014.

27. Clarke, A.; Weeks, I.; Snidle, R.; Evans, H. Running-in and micropitting behaviour of steel surfaces under mixed lubrication conditions. *Tribol. Int.* **2016**, *101*, 59–68. [CrossRef]

28. Jao, T.; Rollin, A.; Carter, R.; Aylott, C.; Shaw, B. Influence of Material Property on Micropitting and Pitting Behavior. In Proceedings of the World Tribology Congress III, Washington, DC, USA, 12–16 September 2005; Volume 2, pp. 103–104.

29. Oila, A.; Bull, S. Phase transformations associated with micropitting in rolling/sliding contacts. *J. Mater. Sci.* **2005**, *40*, 4767–4774. [CrossRef]

30. D'Errico, F. Micropitting Damage Mechanism on Hardened and Tempered, Nitrided, and Carburizing Steels. *Mater. Manuf. Processes* **2011**, *26*, 7–13. [CrossRef]

31. Martins, R.; Seabra, J.; Magalhães, L. Micropitting of Austempered Ductile Iron Gears: Biodegradable Ester vs. Mineral Oil. *Revista da Associação Portuguesa de Análise Experimental de Tensões* **2006**, *122*, 922.

32. Wilkinson, C.; Olver, A. The Durability of Gear and Disc Specimens-Part I: The Effect of Some Novel Materials and Surface Treatments. *ASLE Trans.* **1999**, *42*, 503–510. [CrossRef]

33. Oila, A.; Shaw, B.; Aylott, C.; Bull, S. Martensite decay in micropitted gears. *Proc. Inst. Mech. Eng. Part J J. Eng. Tribol.* **2005**, *219*, 77–83. [CrossRef]

34. Le, M.; Ville, F.; Kleber, X.; Cavoret, J.; Sainte-Catherine, M.; Briancon, L. Influence of grain boundary cementite induced by gas nitriding on the rolling contact fatigue of alloyed steels for gears. *Proc. Inst. Mech. Eng. Part J J. Eng. Tribol.* **2015**, *229*, 917–928. [CrossRef]

35. Tobie, T.; Hippenstiel, F.; Mohrbacher, H. Optimizing gear performance by alloy modification of carburizing steels. *Metals* **2017**, *7*, 415. [CrossRef]

36. Roy, S.; Ooi, G.; Sundararajan, S. Effect of retained austenite on micropitting behavior of carburized AISI 8620 steel under boundary lubrication. *ACTA Mater.* **2018**, *3*, 192–201. [CrossRef]

37. Liu, S.; Song, C.; Zhu, C.; Ni, G. Effects of tooth modifications on mesh characteristics of crossed beveloid gear pair with small shaft angle. *Mech. Mach. Theory* **2018**, *119*, 142–160. [CrossRef]

38. Kissling, U. Application of the first international calculation method for micropitting. *Gear Technol.* **2012**, *29*, 54–60.

39. Predki, W.; Nazifi, K.; Lutzig, G. Micropitting of Big Gearboxes: Influence of Flank Modification and Surface Roughness. *Gear Technol.* **2011**, 42–46.

40. Li, S. An investigation on the influence of misalignment on micro-pitting of a spur gear pair. *Tribol. Lett.* **2015**, *60*, 35. [CrossRef]

41. Weber, C.; Tobie, T.; Stahl, K. Investigation on the flank surface durability of gears with increased pressure angle. *Forsch. Ingenieurwes* **2017**, *81*, 207–213. [CrossRef]

42. Ni, G.; Zhu, C.; Song, C.; Shi, J.; Liu, S. Effects of rack-cutter parabolic modification on loaded contact characteristics for crossed beveloid gears with misalignments. *Int. J. Mech. Sci.* **2018**, *141*, 359–371. [CrossRef]

43. Liu, S.; Song, C.; Zhu, C.; Fan, Q. Concave modifications of tooth surfaces of beveloid gears with crossed axes. *Proc. Inst. Mech. Eng. C J. Mech.* **2018**, 0954406218768842. [CrossRef]

44. Clarke, A.; Evans, H.P.; Snidle, R. Understanding micropitting in gears. *Proc. Inst. Mech. Eng. C J. Mech.* **2016**, *230*, 1276–1289. [CrossRef]

45. Bell, M.; Sroka, G.; Benson, R. The Effect of the Surface Roughness Profile on Micropitting. *Gear Solutions*, 8 March 2013; pp. 46–53.

46. Morales-Espejel, G.; Rycerz, P.; Kadiric, A. Prediction of micropitting damage in gear teeth contacts considering the concurrent effects of surface fatigue and mild wear. *Wear* **2018**, *398*, 99–115. [CrossRef]

47. Evans, H.; Snidle, R.; Sharif, K.; Shaw, B.; Zhang, J. Analysis of micro-elastohydrodynamic lubrication and prediction of surface fatigue damage in micropitting tests on helical gears. *J. Tribol.* **2013**, *135*, 011501. [CrossRef]

48. Sheng, L.; Ahmet, K. A physics-based model to predict micro-pitting lives of lubricated point contacts. *Int. J. Fatigue* **2013**, *47*, 205–215. [CrossRef]

49. Li, S.; Kahraman, A. A micro-pitting model for spur gear contacts. *Int. J. Fatigue* **2014**, *59*, 224–233. [CrossRef]

50. Li, S. A computational study on the influence of surface roughness lay directionality on micropitting of lubricated point contacts. *J. Tribol.* **2015**, *137*, 021401. [CrossRef]

51. AL-Mayali, M.; Hutt, S.; Sharif, K.; Clarke, A.; Evans, H. Experimental and Numerical Study of Micropitting Initiation in Real Rough Surfaces in a Micro-elastohydrodynamic Lubrication Regime. *Tribol. Lett.* **2018**, *66*, 150. [CrossRef]

52. Mallipeddi, D.; Norell, M.; Sosa, M.; Nyborg, L. The effect of manufacturing method and running-in load on the surface integrity of efficiency tested ground, honed and superfinished gears. *Tribol. Int.* **2019**, *131*, 277–287. [CrossRef]

53. Liu, S.; Wang, Q.; Liu, G. A versatile method of discrete convolution and FFT (DC-FFT) for contact analyses. *Wear* **2000**, *243*, 101–111. [CrossRef]

54. Zhang, Y.; Liu, H.; Zhu, C.; Liu, M.; Song, C. Oil film stiffness and damping in an elastohydrodynamic lubrication line contact-vibration. *J. Mech. Sci. Technol.* **2016**, *30*, 3031–3039. [CrossRef]

55. Zhou, Y.; Zhu, C.; Liu, H.; Song, C.; Li, Z. A numerical study on the contact fatigue life of a coated gear pair under EHL. *Ind. Lubr. Tribol.* **2017**, *70*, 23–32. [CrossRef]

56. Liu, H.; Zhu, C.; Sun, Z.; Song, C. Starved lubrication of a spur gear pair. *Tribol. Int.* **2016**, *94*, 52–60. [CrossRef]

57. Charkaluk, E.; Constantinescu, A.; Maïtournam, H.; Van, D. Revisiting the Dang Van criterion. *Procedia Eng.* **2009**, *1*, 143–146. [CrossRef]

58. Osman, T.; Velex, P. A model for the simulation of the interactions between dynamic tooth loads and contact fatigue in spur gears. *Tribol. Int.* **2012**, *46*, 84–96. [CrossRef]

59. Liu, H.; Liu, H.; Zhu, C.; He, H.; Wei, P. Evaluation of Contact Fatigue Life of a Wind Turbine Gear Pair Considering Residual Stress. *J. Tribol.* **2018**, *140*, 041102. [CrossRef]

60. Wang, W.; Liu, H.; Zhu, C.; Du, X.; Tang, J. Effect of the residual stress on contact fatigue of a wind turbine carburized gear with multiaxial fatigue criteria. *Int. J. Mech. Sci.* **2019**, *151*, 263–273. [CrossRef]

61. Hua, Q. Prediction of Contact Fatigue for the Rough Surface Elastohydrodynamic Lubrication Line Contact Problem under Rolling and Sliding Conditions. Ph.D. Thesis, Cardiff University, Wales, UK, 2005.

62. Ciavarella, M.; Maitournam, H. On the Ekberg, Kabo and Andersson calculation of the Dang Van high cycle fatigue limit for rolling contact fatigue. *Fatigue Fract. Eng. Mech.* **2004**, *27*, 523–526. [CrossRef]

63. Liu, H.; Liu, H.; Zhu, C.; Sun, Z.; Wei, P. Study on contact fatigue of a wind turbine gear pair using the EHL model considering surface roughness. *Friction,* (accepted).

64. Karpuschewski, B.; Knoche, H.; Hipke, M. Gear finishing by abrasive processes. *CIRP Ann. Manuf. Technol.* **2008**, *57*, 621–640. [CrossRef]

65. Krantz, T.; Alanou, M.; Evans, H.; Snidle, R. Surface fatigue lives of case-carburized gears with an improved surface finish. *J. Tribol.* **2001**, *123*, 709–716. [CrossRef]

66. Winkelmann, L.; El-Saeed, O.; Bell, M. The effect of superfinishing on gear micropitting. *Gear Technol.* **2009**, *2*, 60–65.

67. Rabaso, P.; Gauthier, T.; Diaby, M.; Ville, F. Rolling Contact Fatigue: Experimental Study of the Influence of Sliding, Load, and Material Properties on the Resistance to Micropitting of Steel Discs. *Tribol. Trans.* **2013**, *56*, 203–214. [CrossRef]

68. Shaikh, J.; Jain, N.; Venkatesh, V. Precision Finishing of Bevel Gears by Electrochemical Honing. *Mater. Manuf. Processes* **2013**, *28*, 1117–1123. [CrossRef]

69. Ronkainen, H.; Elomaa, O.; Varjus, S.; Kilpi, L.; Jaatinen, T.; Koskinen, J. The influence of carbon based coatings and surface finish on the tribological performance in High-load contacts. *Tribol. Int.* **2016**, *96*, 402–409. [CrossRef]

70. Pariente, I.F.; Guagliano, M. Influence of shot peening process on contact fatigue behavior of gears. *Mater. Manuf. Processes* **2009**, *24*, 1436–1441. [CrossRef]

71. Moorthy, V.; Shaw, B. Effect of as-ground surface and the BALINIT® C and Nb–S coatings on contact fatigue damage in gears. *Tribol. Int.* **2012**, *51*, 61–70. [CrossRef]

72. Widmark, M.; Melander, A. Effect of material, heat treatment, grinding and shot peening on contact fatigue life of carburised steels. *Int. J. Fatigue* **1999**, *21*, 309–327. [CrossRef]

73. Terrin, A.; Meneghetti, G. A comparison of rolling contact fatigue behaviour of 17NiCrMo6-4 case-hardened disc specimens and gears. *Fatigue Fract. Eng. Mach.* **2018**, *41*, 2321–2337. [CrossRef]

74. Pariente, I.; Guagliano, M. Contact fatigue damage analysis of shot peened gears by means of X-ray measurements. *Eng. Fail. Anal.* **2009**, *16*, 964–971. [CrossRef]

75. Koenig, J.; Koller, P.; Tobie, T.; Stahl, K. Correlation of relevant case properties and the flank load carrying capacity of case-hardened gears. In Proceedings of the ASME 2015 International Design Engineering Technical Conferences and Computers and Information in Engineering Conference, Boston, MA, USA, 2–5 August 2015.

76. Qin, H.; Ren, Z.; Zhao, J.; Ye, C.; Doll, G.; Dong, Y. Effects of ultrasonic nanocrystal surface modification on the wear and micropitting behavior of bearing steel in boundary lubricated steel-steel contacts. *Wear* **2017**, *392*, 29–38. [CrossRef]

77. Krantz, T.; Cooper, C.; Townsend, D.; Hansen, B. Increased surface fatigue lives of spur gears by application of a coating. *J. Mech. Des.* **2004**, *126*, 1047–1054. [CrossRef]

78. Vetter, J.; Barbezat, G.; Crummenauer, J.; Avissar, J. Surface treatment selections for automotive applications. *Surf. Coat. Technol.* **2005**, *200*, 1962–1968. [CrossRef]

79. Martins, R.; Amaro, R.; Seabra, J. Influence of low friction coatings on the scuffing load capacity and efficiency of gears. *Tribol. Int.* **2008**, *41*, 234–243. [CrossRef]

80. Bayón, R.; Zubizarreta, C.; Nevshupa, R.; Carlos Rodriguez, J.; Fernández, X.; Ruiz de Gopegui, U.; Igartua, A. Rolling-sliding, scuffing and tribocorrosion behaviour of PVD multilayer coatings for gears application. *Ind. Lubr. Tribol.* **2011**, *63*, 17–26. [CrossRef]

81. Moorthy, V.; Shaw, B. Contact fatigue performance of helical gears with surface coatings. *Wear* **2012**, *276–277*, 130–140. [CrossRef]
82. Moorthy, V.; Shaw, B. An observation on the initiation of micro-pitting damage in as-ground and coated gears during contact fatigue. *Wear* **2013**, *297*, 878–884. [CrossRef]
83. Singh, H.; Ramirez, G.; Eryilmaz, O.; Greco, A.; Doll, G.; Erdemir, A. Fatigue resistant carbon coatings for rolling/sliding contacts. *Tribol. Int.* **2016**, *98*, 172–178. [CrossRef]
84. Benedetti, M.; Fontanari, V.; Torresani, E.; Girardi, C.; Giordanino, L. Investigation of lubricated rolling sliding behaviour of WC/C, WC/C-CrN, DLC based coatings and plasma nitriding of steel for possible use in worm gearing. *Wear* **2017**, *378*, 106–113. [CrossRef]
85. Olver, A. The mechanism of rolling contact fatigue: An update. *Proc. Inst. Mech. Eng. J J. Eng. Tribol.* **2005**, *219*, 313–330. [CrossRef]
86. Fajdiga, G.; Flašker, J.; Glodež, S.; Hellen, T. Numerical modelling of micro-pitting of gear teeth flanks. *Fatigue Fract. Eng. Mach.* **2003**, *26*, 1135–1143. [CrossRef]
87. Webster, M.; Norbart, C. An Experimental Investigation of Micropitting Using a Roller Disk Machine. *ASLE Trans.* **1995**, *38*, 883–893. [CrossRef]
88. Moallem, H.; Akbarzadeh, S.; Ariaei, A. Prediction of micropitting life in spur gears operating under mixed-lubrication regime using load-sharing concept. *Proc. Inst. Mech. Eng. J J. Eng. Tribol.* **2016**, *230*, 591–599. [CrossRef]
89. Mallipeddi, D.; Norell, M.; Sosa, M.; Nyborg, L. Influence of running-in on surface characteristics of efficiency tested ground gears. *Tribol. Int.* **2017**, *115*, 45–58. [CrossRef]
90. An, S.; Lee, S.; Son, J.; Cho, Y. New Approach for Prediction of Non-Conformal Contact Fatigue Life Considering Lubrication Performance Parameters. *J. Frict. Wear* **2017**, *38*, 419–423. [CrossRef]
91. Brown, M.; Miller, K. A theory for fatigue failure under multiaxial stress-strain conditions. *Proc. Inst. Mech. Eng.* **1973**, *187*, 745–755. [CrossRef]
92. Kadiric, A.; Rycerz, P. Influence of Contact Conditions on the Onset of Micropitting in Rolling-Sliding Contacts Pertinent to Gear Applications. *Gear Solutions*, 17 February 2017; 45–53.
93. Martins, R.; Seabra, J.; Magalhães, L. Austempered ductile iron (ADI) gears: Power loss, pitting and micropitting. *Wear* **2008**, *264*, 838–849. [CrossRef]
94. Morales-Espejel, G.; Gabelli, A. The Progression of Surface Rolling Contact Fatigue Damage of Rolling Bearings with Artificial Dents. *Tribol. Trans.* **2015**, *58*, 418–431. [CrossRef]
95. Zhou, R.; Cheng, H.; Mura, T. Micropitting in rolling and sliding contact under mixed lubrication. *J. Tribol.* **1989**, *111*, 605–613. [CrossRef]
96. Errichello, R. Morphology of micropitting. *Gear Technol.* **2012**, *4*, 74–81.
97. Sanekata, J.; Koga, N.; Umezawa, O. Effects of Slip Ratio on Damage and Microcracks in Carburized SCM420 Steel under Rolling Contact Fatigue. *Key Eng. Mater.* **2017**, *741*, 94–98. [CrossRef]
98. Cen, H.; Morina, A.; Neville, A. Effect of slide to roll ratio on the micropitting behaviour in rolling-sliding contacts lubricated with ZDDP-containing lubricants. *Tribol. Int.* **2018**, *122*, 210–217. [CrossRef]
99. Flodin, A.; Andersson, S. A simplified model for wear prediction in helical gears. *Wear* **2001**, *249*, 285–292. [CrossRef]
100. Krantz, T.; Kahraman, A. An experimental investigation of the influence of the lubricant viscosity and additives on gear wear. *Tribol. Trans.* **2005**, *21*, 138–148. [CrossRef]
101. Brandão, J.; Martins, R.; Seabra, J.; Castro, M. Calculation of gear tooth flank surface wear during an FZG micropitting test. *Wear* **2014**, *311*, 31–39. [CrossRef]
102. Al-Mayali, M.; Evans, H.; Sharif, K. Assessment of the effects of residual stresses on fatigue life of real rough surfaces in lubricated contact. In Proceedings of the International Conference for Students on Applied Engineering, Newcastle, UK, 20–21 October 2016; pp. 123–128.
103. Janakiraman, V. An Investigation of the Impact of Contact Parameters on the Wear Coefficient. Ph.D. Thesis, The Ohio State University, Columbus, OH, USA, 2013.
104. Liu, H.; Liu, H.; Zhu, C.; Tang, J. Study on contact fatigue failure competitive mechanism of a wind turbine gear pair considering tooth wear evolution. *J. Tribol.* (under review).
105. Winter, H.; Oster, P. Influence of lubrication on pitting and micropitting resistance of gears. *Gear Technol.* **1990**, *7*, 16–23.

106. Morales-Espejel, G.; Brizmer, V.; Piras, E. Roughness evolution in mixed lubrication condition due to mild wear. *Proc. Inst. Mech. Eng. J J. Eng. Tribol.* **2015**, *229*, 1330–1346. [CrossRef]

107. Vrcek, A.; Hultqvist, T.; Baubet, Y.; Björling, M.; Marklund, P.; Larsson, R. Micro-pitting and wear assessment of engine oils operating under boundary lubrication conditions. *Tribol. Int.* **2019**, *129*, 338–346. [CrossRef]

108. Lainé, E.; Olver, A.; Beveridge, T. Effect of lubricants on micropitting and wear. *Tribol. Int.* **2008**, *41*, 1049–1055. [CrossRef]

109. Van-Rensselar, J. Trends in industrial gear oils. *Tribol. Lubr. Technol.* **2013**, *69*, 26–33.

110. Errichello, R. Selecting and applying lubricants to avoid micropitting of gear teeth. *Mach. Lubr.* **2002**, *2*, 30–36.

111. Martins, R.; Seabra, J. Micropitting performance of mineral and biodegradable ester gear oils. *Ind. Lubr. Tribol.* **2008**, *60*, 286–292. [CrossRef]

112. Cardoso, R.; Martins, C.; Seabra, O.; Igartua, A.; Rodríguez, C.; Luther, R. Micropitting performance of nitrided steel gears lubricated with mineral and ester oils. *Tribol. Int.* **2009**, *42*, 77–87. [CrossRef]

113. Lainé, E.; Olver, A.; Lekstrom, M.; Shollock, B.; Beveridge, T.; Hua, D. The Effect of a Friction Modifier Additive on Micropitting. *Tribol. Trans.* **2009**, *52*, 526–533. [CrossRef]

114. De la Guerra Ochoa, E.; Otero, J.E.; Tanarro, E.C.; Munoz-Guijosa, J.; del Rio Lopez, B.; Cordero, C.A. Analysis of the effect of different types of additives added to a low viscosity polyalphaolefin base on micropitting. *Wear* **2015**, *322*, 238–250. [CrossRef]

115. Soltanahmadi, S.; Morina, A.; van Eijk, M.C.; Nedelcu, I.; Neville, A. Investigation of the effect of a diamine-based friction modifier on micropitting and the properties of tribofilms in rolling-sliding contacts. *J. Phys. D Appl. Phys.* **2016**, *49*, 505302. [CrossRef]

116. Engelhardt, C.; Witzig, J.; Tobie, T.; Stahl, K. Influence of water contamination in gear lubricants on wear and micro-pitting performance of case carburized gears. *Ind. Lubr. Tribol.* **2017**, *69*, 612–619. [CrossRef]

117. Ward, W.; O'connor, B.; Vinci, J. Lubricants That Decrease Micropitting for Industrial Gears. U.S. Patent Application 11/866,696, 9 April 2009.

118. Fu, X.; Hua, X.; Zhang, J. Industrial Gear Lubricating Oil Composition Used for Resisting Micro-Pitting. U.S. Patent 9,347,016, 24 May 2016.

119. Moss, J.; Kahraman, A.; Wink, C. An Experimental Study of Influence of Lubrication Methods on Efficiency and Contact Fatigue Life of Spur Gears. *J. Tribol.* **2018**, *140*, 051103. [CrossRef]

120. Seireg, A. Thermal stress effects on the surface durability of gear teeth. *Proc. Inst. Mech. Eng. C J. Mech.* **2001**, *215*, 973–979. [CrossRef]

121. Zhou, Y.; Zhu, C.; Gould, B.; Demas, N.; Liu, H.; Greco, A. The effect of contact severity on micropitting: simulation and experiments. *Tribol. Int.* (under review).

122. Morales-Espejel, G.; Brizmer, V. Micropitting Modelling in Rolling–Sliding Contacts: Application to Rolling Bearings. *Tribol. Trans.* **2011**, *54*, 625–643. [CrossRef]

123. Archard, F. Contact and rubbing of flat surfaces. *J. Appl. Phys.* **1953**, *24*, 981–988. [CrossRef]

124. Brizmer, V.; Pasaribu, H.; Morales-Espejel, G. Micropitting Performance of Oil Additives in Lubricated Rolling Contacts. *Tribol. Trans.* **2013**, *56*, 739–748. [CrossRef]

125. Benyajati, C.; Olver, A.; Hamer, C. An experimental study of micropitting using a new miniature test-rig. *Tribol. Ser.* **2003**, *43*, 601–610.

126. O'connor, B. The influence of additive chemistry on micropitting. *Gear Technol.* **2005**, *22*, 34–41.

127. Yang, Q. Fatigue test and reliability design of gears. *Int. J. Fatigue* **1996**, *18*, 171–177. [CrossRef]

128. Li, F.; Hu, W.; Meng, Q.; Zhan, Z.; Shen, F. A new damage-mechanics-based model for rolling contact fatigue analysis of cylindrical roller bearing. *Tribol. Int.* **2017**, *120*, 105–114. [CrossRef]

129. Weibring, M.; Gondecki, L.; Tenberge, P. Simulation of fatigue failure on tooth flanks in consideration of pitting initiation and growth. *Tribol. Int.* **2019**, *131*, 299–307. [CrossRef]

130. Wang, W.; Liu, H.; Zhu, C.; Wei, P.; Tang, J. Effects of microstructure on rolling contact fatigue of a wind turbine gear based on crystal plasticity modeling. *Int. J. Fatigue* **2019**, *120*, 73–86. [CrossRef]

131. Rajinikanth, V.; Soni, M.K.; Mahato, B.; Rao, M.A. Study of microstructural degradation of a failed pinion gear at a cement plant. *Eng. Fail. Anal.* **2019**, *95*, 117–126. [CrossRef]

coatings

MDPI

Article

Microstructure and Cavitation Erosion Resistance of HVOF Deposited WC-Co Coatings with Different Sized WC

Xiang Ding [1], Du Ke [2], Chengqing Yuan [2,*], Zhangxiong Ding [2] and Xudong Cheng [1]

[1] State Key Laboratory of Advanced Technology for Materials and Processing, Wuhan University of Technology, Wuhan 430070, China; dingxiang@whut.edu.cn (X.D.); xudong.cheng@whut.edu.cn (X.C.)
[2] School of Energy and Power Engineering, Wuhan University of Technology, Wuhan 430063, China; kedu_t@whut.edu.cn (D.K.); zxding@whut.edu.cn (Z.D.)
* Correspondence: ycq@whut.edu.cn; Tel.: +86-27-8658-2035

Received: 25 July 2018; Accepted: 24 August 2018; Published: 29 August 2018

Abstract: Conventional, multimodal and nanostructured WC-12Co coatings with different WC sizes and distributions were prepared by high velocity oxy-fuel spray (HVOF). The micrographs and structures of the coatings were analyzed by scanning electron microscope (SEM), X-ray diffractometer (XRD) et al. The porosity, microhardness and fracture toughness of the WC-Co coatings were measured. The coating resistance to cavitation erosion (CE) was investigated by ultrasonic vibration cavitation equipment and the cavitation mechanisms were explored. Results show that there is serious WC decarburization in nanostructured and multimodal WC-Co coatings with the formation of W_2C and W phases. The nanostructured WC-Co coating has the densest microstructure with lowest porosity compared to the other two WC-Co coatings, as well as the highest fracture toughness among the three coatings. It was also discovered that the nanostructured WC-Co coating exhibits the best CE resistance and that the CE rate is approximately one-third in comparison with conventional coating.

Keywords: WC-12Co coatings; HVOF; microstructure; cavitation erosion

1. Introduction

Cavitation erosion (CE) widely exists in the components of fluid equipment, such as marine rudder blades, propellers, and turbine impellers. CE is the predominant cause for overflow part failure and has become one of the most significant technical problems of fluid machinery due to the serious threats to the safety of the machinery, leading to the reduction of the efficiency and the increase of the production cost [1–3]. Therefore, it is of great economic value to improve the CE resistance of the overcurrent components. Since cavitation occurs only in the components' surface, the CE resistance can be improved by surface engineering techniques. Nowadays, various surface engineering techniques, including thermal spraying, laser cladding, physical vapor deposition, chemical vapor deposition, hardening and plasma nitriding, have been developed to enhance the CE resistance of component surfaces. Among them, thermal spraying methods, such as detonation gun, plasma spraying, high velocity oxy-fuel spraying and cold spraying, have already been commercially applied to various machinery components [4–7]. Thermally sprayed WC-Co coatings have attracted much attention in the study of CE resistant coating materials in recent years due to their combination of excellent toughness and high hardness, which is necessary to CE resistance. WC particles and Co binders provide WC-Co coatings with high hardness and excellent toughness, respectively. WC-Co coatings, especially with nano-sized WC particles, have already been successfully utilized in some wear-resistant equipment [8,9].

The mechanical properties and wear performance of WC-Co coatings are largely dependent on WC grain size [10,11]. Recently, simultaneous improvement of the toughness and hardness of nanostructured WC-Co coatings were successfully demonstrated by a number of studies [12–15]. However, the issue of how exactly nano WC affect the wear performance has been a controversial and disputed subject due to various coating fabrication methods and parameters [16,17]. Several researchers have also discovered that the fracture toughness of WC-Co coatings decreases as the WC particle size decreases because of the decarburization of nano WC and the formation of amorphous carbides such as W_2C, complex Co-W-C, and metallic W, which can lower the toughness of nanostructured WC-Co coatings [18–20]. The multimodal WC-Co coating composed of nano and micro WC particles, on the other hand, presents lower decarburization and total cost as compared to the nano WC-Co coatings [21–23]. Compared with the conventional WC-Co coatings, multimodal WC coatings offer a denser structure, higher abrasive wear resistance and anti-cavitation performance [24–26]. For example, by comparing the microstructure and surface properties of conventional and nano WC-Co coatings, Zhao et al. [27] concluded lower porosity, higher microhardness and fracture toughness could result from the nano coatings. The microstructure and CE resistance of other multimodal coatings, such as WC-17Co and WC-10Co4Cr have been further investigated by a number of groups [28,29]. However, the relationship of structures, mechanical properties, CE resistance and mechanisms of WC-Co coatings with different WC size have not been studied in details and fully understood.

The coating deposition method is another critical factor influencing the structures and properties of WC-Co cermet coatings. To date, WC-Co coatings are most commonly prepared by HVOF technique because HVOF flame possesses the characteristics of moderate temperature (~2700 °C) and high velocity (600–1200 m/s), which can reduce the decarburization of WC during coating deposition. Therefore, HVOF is thought to be a proper process to deposit WC-Co coatings with nano sized WC particles [30,31].

In the present paper, conventional, multimodal and nanostructured WC-12Co cermet coatings were prepared by HVOF spraying. The structures, mechanical properties and CE performance of the three types of WC-12Co coatings were thoroughly investigated. The CE mechanisms of the coatings were proposed in terms of the formation and propagation of microcracks. These results can provide valuable references for WC-Co anti-cavitation coating design and application.

2. Experimental Procedure

2.1. Coating Materials

Nanostructured, conventional and multimodal WC-12Co (88 wt % WC-12 wt % Co) powders were chosen as feedstock in this study and they were marked as NP, CP, and MP, respectively. Both nanostructured WC-12Co powder (S7410, Inframat Advanced Materials LLC, Farmington, CT, USA) and multimodal powder (M1, NEI, Somerset, NJ, USA) were produced by an agglomeration process. The original WC grain size of NP is 50–500 nm, while the volume ratio of micro-sized WC grain (2–3 μm) and nano-sized WC grain (30–50 nm) in MP is 7:3. The size of obtained spraying powder is controlled in the range of 10–45 μm and the mean particle size is 23 μm after agglomeration. On the other hand, the conventional WC-12Co powder (PR4321, Jiorie Thermal Spray Materials Co., Yiyang, China), with a size also between 10 and 45 μm, was fabricated by sintering and crushing process, and the original WC grain size was 2–3 μm.

2.2. Coating Fabrication

The conventional, multimodal and nanostructured WC-12Co coatings were fabricated by a HVOF system (JP5000, Praxair, Inc., Indianapolis, IN, USA) and they were marked as CC, MC and NC, respectively. In the spray system, kerosene and oxygen were used as fuel and oxidant gas. The optimized main spray parameters are shown in Table 1. Due to the variations in particle velocities and temperature characteristics during HVOF spraying for various sized WC-Co materials, deposition parameters are different for three WC-12Co powders. AISI 304 austenitic stainless steel was chosen as the substrate of coating specimens and as the reference material for comparing the CE resistance of HVOF-sprayed various sized WC-12Co coatings.

Table 1. Main spray parameters of WC-12Co coatings by high velocity oxy-fuel spray (HVOF). NP: nanostructured powder; MP: multimodal powder; CP: conventional powder; NC: nanostructured coating; MC: multimodal coating; CC: conventional coating.

Powder No.	Coating No.	Fuel Flow (m^3/h)	Oxygen Flow (m^3/h)	Powder Feed Rate (g/min)	Spray Distance (mm)	Horizontal Velocity (mm/s)	Vertical Step (mm)
NP	NC	0.019	53.2	75	370	500	5
MP	MC	0.019	53.2	75	370	500	5
CP	CC	0.0204	61.3	75	380	500	5

Prior to deposition, the substrates' surface was degreased and grit-blasted with 60 mesh Al_2O_3. The coating thickness was in the range of 400 ± 20 μm. All the specimens were machined and polished to surface roughness $R_a \leq 0.02$ μm for coating characterization.

2.3. Characterization

An X-ray diffraction system (XRD, D/max-2550 diffraction meter, Rigaku Corporation, Tokyo, Japan) was used for phase identification of the WC-12Co powders and the coatings. The operation of the Cu Kα radiation source was under a voltage of 45 kV and a current of 40 mA. JSM6700F scanning electron microscope (SEM) was carried out to observe the surface morphology and microstructure of the samples. The microhardness of the coatings was measured by a Vickers microhardness tester (Model HV-71, Aolong Xingdi Testing Equipment Co., Shanghai, China). The final value was determined by the average of random ten test points at a loading weight of 200 g and a dwell time of 15 s. The porosity was evaluated by two steps. $500\times$ metallographic photos were first captured by Leitz MM6 metallographic microscope (Leica, Wetzlar, Germany), followed by applying IQ materials software (version 2.0) to calculate porosity. According to [32], fracture toughness can be measured on the transverse section of the coating with a Vickers indenter (HV-5 Vickers hardness tester, 5 kg). Cracks parallel to the substrate surface such as those drawn in Figure 1 would appear on the coating's cross-section. The fracture toughness K_c was calculated according to the Wilshaw Equation (1) and the result was the average value of ten measurements, where P, a and c are the load of hardness meter, half the length of indentation diagonal and half the length of the crack, respectively.

$$K_C = 0.079 \frac{P}{a^{3/2}} \log\left(\frac{4.5a}{c}\right) \tag{1}$$

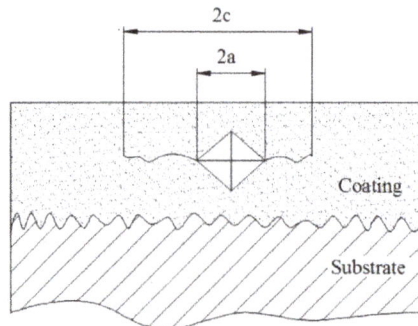

Figure 1. Diagram of coating fracture toughness measurement.

2.4. Cavitation Erosion Tests

A J93025 ultrasonic cavitation vibratory equipment was used to evaluate the cavitation erosion of various WC-12Co coatings. The apparatus was set up according to standard GB/T6383-2009 [33], and the schematic diagram is shown in Figure 2. The specimen was machined to the surface roughness $R_a \leq 0.2$ µm before CE tests and then attached to the end of the vibration probe. The samples were weighed every hour by AB204-S electronic balance (Mettler-Toledo GmbH, Greifensee, Switzerland) with the accuracy of 0.1 mg to determine the mass losses during the whole 16-h-test. The volume loss (ΔV) is calculated by mass loss (ΔW) divided by the material density, while the cavitation rate (R_c) is calculated by volume loss per hour. In addition, the CE tests of the austenitic stainless steel AISI 304 samples were performed under the same test conditions for comparison.

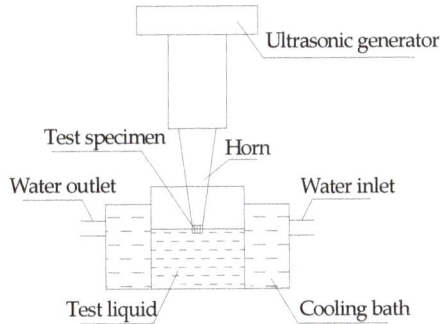

Figure 2. Schematic diagram of cavitation erosion (CE) test apparatus.

3. Results and Discussion

3.1. Microstructure of WC-12Co Powders and Coatings

The surface micrographs of three WC-12Co powders are present in Figure 3. NP and MP demonstrate spherical shape, but CP shape shows more corners and edges. From CP to MP and NP powder, the WC grains decrease in size meanwhile distribute more evenly. In NP powder, the original crystal WC size is less than 100 nm. The largest grains in CP powder reach 2–3 µm, but a small amount of WC particles remain with a sub-micron size.

The surface morphology of three WC-12Co coatings sprayed by HVOF with different WC sizes are shown in Figure 4. Polygon-shaped, unmolten WC grains can be observed in the CC coating. This suggests that HVOF flame only melts the Co binder, while WC is still in a solid state. In NC and MC coatings, most nano-WC grains are dissolved by heating due to their finer size and higher surface to volume ratio. Among the three WC-Co coatings, NC shows the best melting condition. Thus, the coating microstructures are strongly dependent on the powder structures.

Figure 5 presents the XRD patterns of WC-12Co powders and coatings deposited by HVOF with different WC sizes. The XRD patterns of NP, MP and CP powders show no differences, consisting of pure WC and Co, as shown in the CP pattern. The CC coating has almost identical phase compositions to CP powder, which are mainly composed of WC and Co. For MC and NC coatings, the coatings consist of WC, W_2C, and W crystalline phases; the latter two are generated by the decarburization of nano WC. Also, it is demonstrated that an amorphous nanocrystalline zone exists between 35° and 48° at 2θ angles in MC and NC coatings. Although both MP and NP powders suffer typical decarburization during coating deposition, the nanostructured WC-12Co powder had a more serious decarburization rate as more metallic W can be observed in Figure 5. It can be concluded that the decreased size of WC grains is the cause for more serious decarburization, because the larger surface contact area with the flame will give the particles more heat to reach higher temperatures. Figure 6 shows cross-sectional microstructures of three WC-12Co coatings. It can be observed in Figure 6f that the amount of carbides in the NC coating decreased considerably, and more metal phases appeared, which is consistent with the above XRD results.

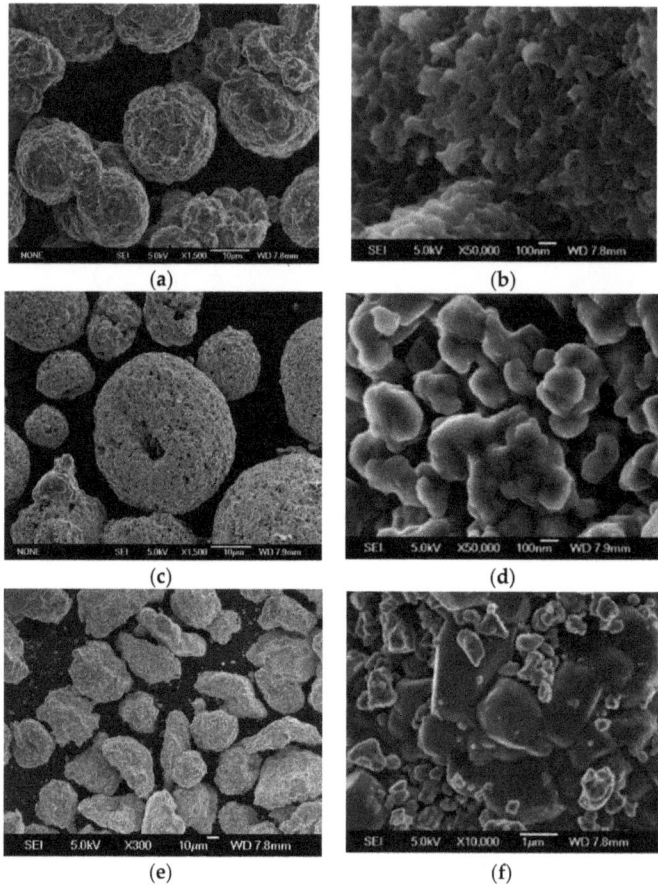

Figure 3. SEM micrographs of (**a,b**) nanostructured (NP), (**c,d**) multimodal (MP), (**e,f**) conventional (CP) WC-12Co powders.

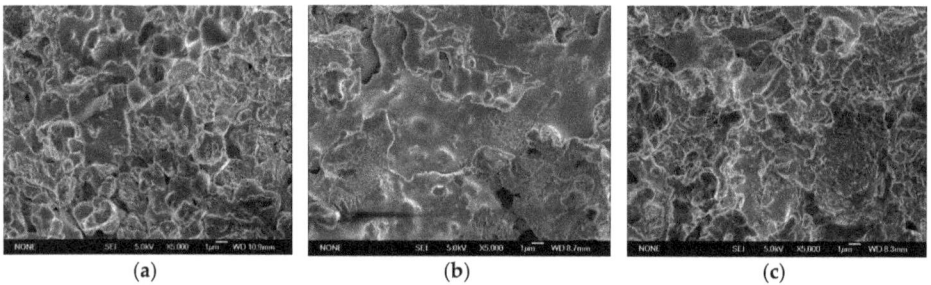

Figure 4. Surface micrographs of different structured WC-12Co coatings: (**a**) CC; (**b**) NC; (**c**) MC.

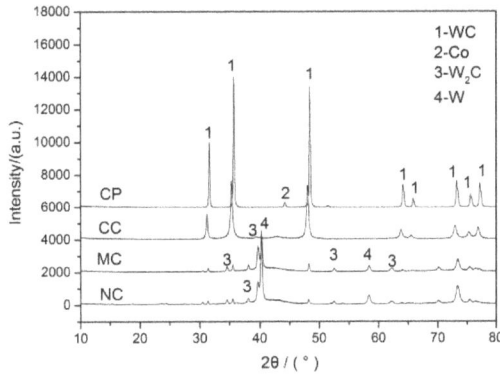

Figure 5. XRD patterns of differently structured WC-12Co powders and coatings.

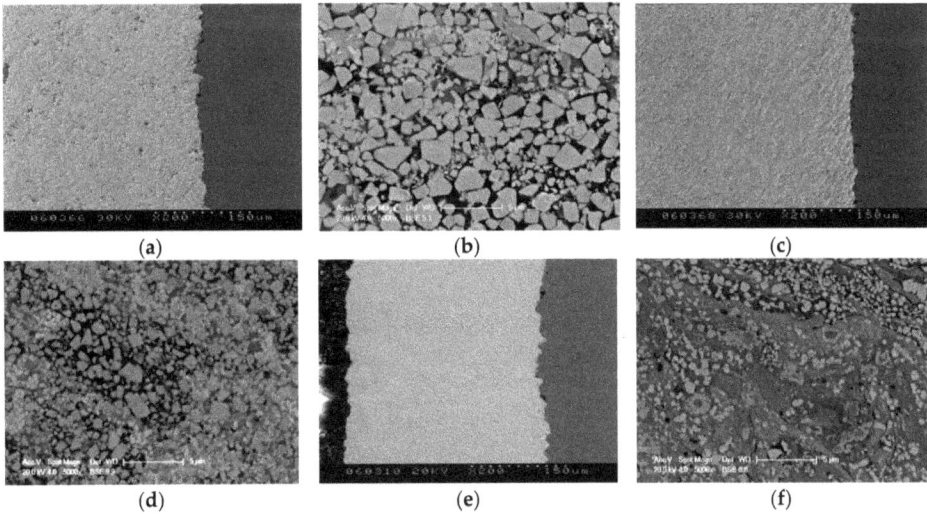

Figure 6. Cross-sectional micrographs of different structured WC-12Co coatings: (**a**,**b**) CC; (**c**,**d**) MC; (**e**,**f**) NC.

It can also be observed in Figure 6a,c,e that the microstructures of the three WC-Co coatings are dense and the interlamellar cohesion is strong. The microstructure of NC is densest in comparison with CC and MC coatings. The porosity of CC, MC and NC coatings sprayed by HVOF is shown in Figure 7. It can be seen that NC coating possesses the lowest porosity (0.63% ± 0.11%), at only 36% compared with CC coating (1.76% ± 0.27%). MC coating shows a moderate porosity (1.18 ± 0.21). This indicates that the nanostructured WC-12Co particles possessed more enthalpy and better melting condition before reaching the substrate, given rise to sufficient deformation and growth of denser structure coating.

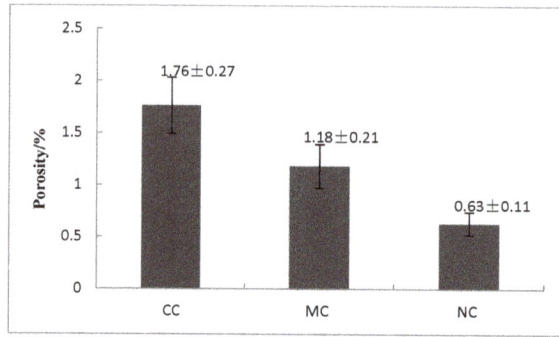

Figure 7. Porosity of different structured WC-12Co coatings.

3.2. Mechanical Properties of WC-12Co Coatings

The mechanical properties, including the fracture toughness and microhardness of three different structured WC-Co coatings, are shown in Table 2. The micrographs of fracture toughness indentation of the coatings are presented in Figure 8. It can be seen from Table 2 that the microhardness values of NC and MC coatings are obviously higher than CC coating, as the nano WC particle size decreases and nano WC content increases. The average microhardness values of NC and MC coatings exceed 1500 $HV_{0.2}$ and are 50% higher than the CC coating. It is also observed that the microhardness values of the MC coating change in a larger range than NC coating, although the average microhardness values of NC and MC coatings are almost the same, because WC grain size in MC coating is more disparate, which would influence the coating microhardness variation. Meanwhile, Table 2 demonstrates that the fracture toughness of NC and MC coatings are higher than that of the CC coating, and the highest value is obtained by NC coating. This may be caused by the W_2C and W phase generated during spraying process, in which nano-sized W_2C particles enhance the coating hardness and metallic W phase improves the coating toughness.

Table 2. Mechanical properties of different structured WC-12Co coatings.

Coating No.	Microhardness ($HV_{0.2}$)	Fracture Toughness (MPa·m$^{1/2}$)
CC	1034 ± 77.5	3.76 ± 0.38
MC	1523 ± 157.0	4.19 ± 0.65
NC	1541 ± 80.0	4.88 ± 0.47

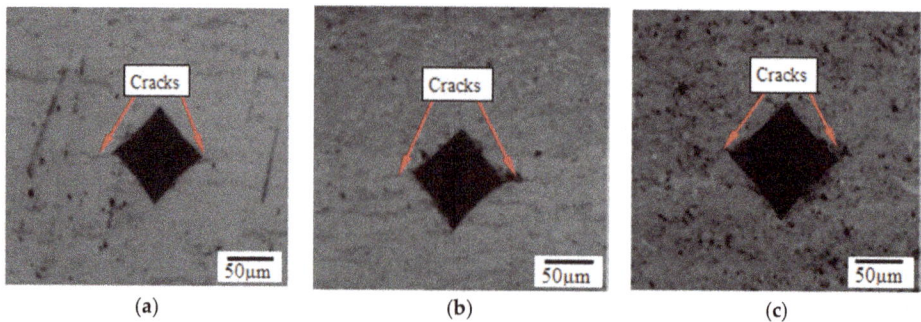

Figure 8. Micrographs of fracture toughness indentation of WC-12Co coatings: (**a**) NC; (**b**) MC; (**c**) CC.

3.3. Cavitation Erosion Resistance

The cumulative volume loss curves of 304 stainless steel and three differently structured WC-12Co coatings sprayed by HVOF in fresh water are illustrated in Figure 9. It can be discovered that three WC-Co coatings all possess superior CE resistance than 304 stainless steel after 16 h cavitation, and NC coating displays the best CE resistance increasing more than 60% and 45% more, respectively, than CC and MC coatings.

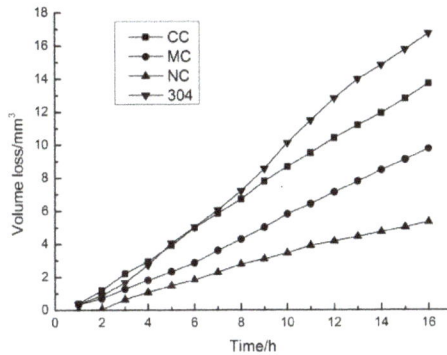

Figure 9. Volume loss curves of WC-12Co coatings.

CE rates of 304 stainless steel and three WC-12Co coatings are shown in Figure 10. It can be demonstrated that, in the steady period, the CE rate of CC coating is approximately 1.0 mm^3/h, while that of NC coating is 0.3 mm^3/h; only 30% of the former. Meanwhile, the CE rate of the MC coating ranks in the middle, at about 70% of the CC coating. It is also seen that in the last testing period (12–16 h), the CE rate of the NC coating displays the tendency to decline compared with the CC coating. From the CE results of Figures 9 and 10, it can be revealed that NC coating with a nano WC size possesses the best CE resistance, while MC coating exhibits moderate CE performance among the three WC-Co coatings.

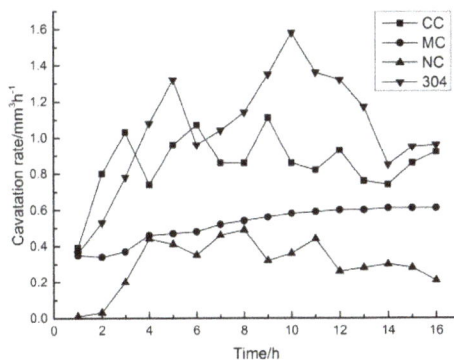

Figure 10. Cavitation rates of WC-12Co coatings.

3.4. Cavitation Erosion Mechanisms

SEM images of the eroded surfaces of three differently structured WC-12Co coatings after 16 h of CE testing are demonstrated in Figure 11. Due to different CE mechanisms, various eroded surface micrographs can be observed from different WC-Co coatings. For the CC coating, microcracks initially originate at the weak points on the surface, such as defects, because of the stress concentration under strong alternative stress from the bubble collapsing. The cracks then propagate along the direction of the microjet at the grain boundary, leading to the breakage of brittle phases at the grain boundary and the formation of a cavitation source. In this case, grain can be easily stripped, forming a crater on the coating surface. The growth of craters can further lead coarse WC particles to be removed and lamellar structures to delaminate (Figure 11a).

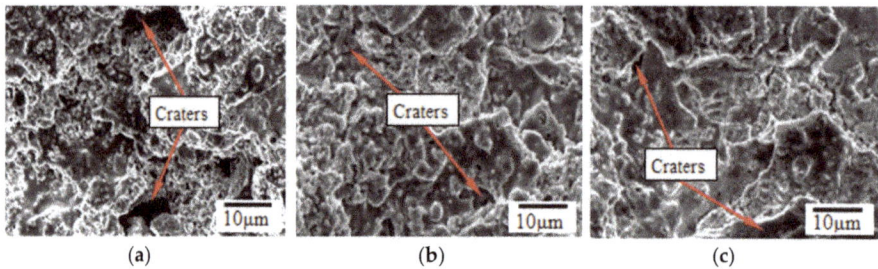

Figure 11. Crater micrographs of WC-12Co coatings after 16 h CE: (**a**) CC; (**b**) MC; (**c**) NC.

NC and MC coatings possess a finer and denser microstructure compared to the CC coating since they are composed of nano and sub-micro particles with a sufficient deformation, which diminish the lamellar structure to a certain extent and enhance the cohesive strength between particles. The large area of the particle interface and uniformly distributed grains create a large amount of nanosized grain boundaries, leading to the increase of the coatings' cohesive strength of the grain boundary. Meanwhile, fine particles with superior microhardness such as W_2C can increase the coatings' microhardness. Moreover, the certain amount of W phase with high toughness raises the coatings' fracture toughness. Thus, NC and MC coatings exhibit better CE resistance because the generation and propagation of microcracks are both hindered and only shallow eroded pits can form, as shown in Figure 11b,c. The best CE resistance is realized by the nanostructured WC-12Co coating, which has the highest microhardness and fracture toughness due to the coexistence of nano-sized WC and W_2C particles and tough Co and W metals.

4. Conclusions

- WC decarburization occurs during HVOF spraying nanostructured and multimodal WC-12Co powders by forming W_2C and W phases, while no apparent decarburization can be observed using conventional WC-12Co powder. Nanostructured WC-12Co coating suffers the most serious decarburization but possesses the lowest porosity.
- Average microhardness values of nanostructured and multimodal WC-12Co coatings exceed 1500 $HV_{0.2}$, which is 50% higher than the conventional coating. Both coatings also show higher fracture toughness, especially the nanostructured coating.
- Nanostructured and multimodal WC-12Co coatings exhibit the best and intermediate CE resistance, respectively, and the CE rate of nanostructured coating is only approximately one third of that of conventional coating.
- The enhanced CE resistance of the nanostructured WC-12Co coating originates from its superb fracture toughness and microhardness, which makes the microcracks form and propagate with more difficulty. Meanwhile, the coatings dense nanostructure and strong cohesive strength are also factors for the excellent cavitation erosion performance.

Author Contributions: Conceptualization, X.D. and Z.D.; Methodology, X.D.; Software, X.D. and D.K.; Validation, X.D., D.K. and C.Y.; Formal Analysis, C.Y.; Investigation, Z.D.; Resources, X.C.; Data Curation, D.K.; Writing—Original Draft Preparation, X.D.; Writing—Review & Editing, X.D. and D.K.; Visualization, Z.D. and X.C.; Supervision, C.Y.; Project Administration, C.Y.; Funding Acquisition, C.Y.

Funding: This research was funded by the National Natural Science Foundation of China (No. 51422507).

Conflicts of Interest: The authors declare no conflict of interest.

References

1. Singh, R.; Tiwari, S.K.; Mishra, S.K. Cavitation erosion in hydraulic turbine components and mitigation by coatings: Current status and future needs. *J. Mater. Eng. Perform.* **2012**, *21*, 1539–1551. [CrossRef]

2. Jasionowski, R.; Zasada, D.; Przetakiewicz, W. Cavitation erosion resistance of alloys used in cathodic protection of hulls of ships. *Arch. Metall. Mater.* **2014**, *59*, 241–245. [CrossRef]

3. Kim, J.H.; Yang, H.S.; Baik, K.H.; Seong, B.G.; Lee, C.H.; Hwang, S.Y. Development and properties of nanostructured thermal spray coatings. *Curr. Appl. Phys.* **2006**, *6*, 1002–1006. [CrossRef]

4. Kamdi, Z.; Shipway, P.H.; Voisey, K.T.; Sturgeon, A.J. Abrasive wear behavior of conventional and large-particle tungsten carbide-based cermet coatings as a function of abrasive size and type. *Wear* **2011**, *271*, 1264–1272. [CrossRef]

5. Tillmann, W.; Baumann, I.; Hollingsworth, P.S.; Hagen, L. Sliding and rolling wear behavior of HVOF-sprayed coatings derived from conventional, fine and nanostructured WC-12Co powders. *J. Therm. Spray Technol.* **2014**, *23*, 262–280. [CrossRef]

6. Krebs, S.; Grtner, F.; Klassen, T. Cold spraying of Cu-Al-Bronze for cavitation protection in marine environments. *J. Therm. Spray Technol.* **2015**, *45*, 708–716.

7. Wang, Y.; Liu, J.; Kang, N.; Darut, G.; Poirier, T.; Stella, J.; Liao, H.; Planche, M.H. Cavitation erosion of plasma-sprayed CoMoCrSi coatings. *Tribol. Int.* **2016**, *102*, 429–435. [CrossRef]

8. Ma, N.; Guo, L.; Cheng, Z.; Wu, H.; Ye, F.; Zhang, K. Improvement on mechanical properties and wear resistance of HVOF sprayed WC-12Co coatings by optimizing feedstock structure. *Appl. Surf. Sci.* **2014**, *320*, 364–371. [CrossRef]

9. Guilemany, J.M.; Dosta, S.; Miguel, J.R. The enhancement of the properties of WC-Co HVOF coatings through the use of nanostructured and microstructured feedstock powders. *Surf. Coat. Technol.* **2006**, *201*, 1180–1190. [CrossRef]

10. Lekatou, A.; Sioulas, D.; Karantzalis, A.E.; Grimanelis, D. A comparative study on the microstructure and surface property evaluation of coatings produced from nanostructured and conventional WC-Co powders HVOF-sprayed on Al7075. *Surf. Coat. Technol.* **2015**, *276*, 539–556. [CrossRef]

11. Babua, P.S.; Basub, B.; Sundararajan, G. Abrasive wear behavior of detonation sprayed WC-12Co coatings: Influence of decarburization and abrasive characteristics. *Wear* **2010**, *268*, 1387–1399. [CrossRef]

12. Cho, T.Y.; Yoon, J.H.; Kim, K.S.; Song, K.O.; Joo, Y.K.; Fang, W.; Zhang, S.H.; Youn, S.J.; Chun, H.G.; Hwang, S.Y. A study on HVOF coatings of micron and nano WC-Co powders. *Surf. Coat. Technol.* **2008**, *202*, 5556–5559. [CrossRef]

13. Yang, Q.; Senda, T.; Ohmori, A. Effect of carbide grain size on microstructure and sliding wear behavior of HVOF-sprayed WC-12%Co coatings. *Wear* **2003**, *254*, 23–34. [CrossRef]

14. He, J.; Lavernia, E.J.; Liu, Y.; Qiao, Y.; Fischer, T.E. Near-nanostructured WC-18 pct Co coatings with low amounts of non-WC carbide phase: Part I. synthesis and characterization. *Metall. Mater. Trans. A* **2002**, *33*, 145–157. [CrossRef]

15. Ji, G.C.; Wang, H.T.; Chen, X.; Bai, X.B.; Dong, Z.X.; Yang, F.G. Characterization of cold-sprayed multimodal WC-12Co coating. *Surf. Coat. Technol.* **2013**, *235*, 536–543. [CrossRef]

16. Wang, Q.; Chen, Z.H.; Li, L.X.; Yang, G.B. The parameters optimization and abrasion wear mechanism of liquid fuel HVOF sprayed bimodal WC-12Co coating. *Surf. Coat. Technol.* **2012**, *206*, 2233–2241. [CrossRef]

17. Ding, Z.X.; Chen, W.; Wang, Q. Resistance of cavitation erosion of multimodal WC-12Co coatings sprayed by HVOF. *Trans. Nonferrous Met. Soc. China* **2011**, *21*, 2231–2236. [CrossRef]

18. Ding, X.; Cheng, X.D.; Shi, J.; Li, C.; Yuan, C.Q.; Ding, Z.X. Influence of WC size and HVOF process on erosion wear performance of WC-10Co4Cr coatings. *Int. J. Adv. Manuf. Technol.* **2018**, *96*, 1615–1624. [CrossRef]

19. Hu, Y.M. Structure and Resistance of Cavitation Erosion Micro-Nano WC-Based Coatings Sprayed by HVOF. Master's Thesis, Wuhan University of Technology, Wuhan, Hubei, China, 2015. (In Chinese)

20. Ghabchi, A.; Varis, T.; Turunen, E.; Suhonen, T.; Liu, X.; Hannula, S.P. Behavior of HVOF WC-10Co4Cr coatings with different carbide size in fine and coarse particle abrasion. *J. Therm. Spray Technol.* **2010**, *19*, 368–377. [CrossRef]

21. Wang, H.T.; Ji, G.C.; Chen, Q.Y.; Du, X.F.; Fu, W. Microstructure characterization and abrasive wear performance of HVOF sprayed WC-Co coatings. *Adv. Mater. Res.* **2011**, *189–193*, 707–710. [CrossRef]

22. Hong, S.; Wu, Y.P.; Zhang, J.F.; Zheng, Y.G.; Zheng, Y.; Lin, J.R. Synergistic effect of ultrasonic cavitation erosion and corrosion of WC-CoCr and FeCrSiBMn coatings prepared by HVOF spraying. *Ultrason. Sonochem.* **2016**, *31*, 563–569. [CrossRef] [PubMed]

23. Wang, Q.; Tang, Z.X.; Cha, L.M. Cavitation and sand slurry erosion resistances of WC-10Co-4Cr coatings. *J. Mater. Eng. Perform.* **2015**, *24*, 2435–2443. [CrossRef]

24. Rodríguez, M.A.; Gil, L.; Camero, S.; Fréty, N.; Santana, Y.; Caro, J. Effects of the dispersion time on the microstructure and wear resistance of WC/Co-CNTs HVOF sprayed coatings. *Surf. Coat. Technol.* **2014**, *258*, 38–48. [CrossRef]

25. Ding, X.; Cheng, X.D.; Li, C.; Yu, X.; Ding, Z.X.; Yuan, C.Q. Microstructure and performance of multi-dimensional WC-CoCr coating sprayed by HVOF. *Int. J. Adv. Manuf. Technol.* **2018**, *96*, 1625–1633. [CrossRef]

26. Hong, S.; Wu, Y.P.; Zhang, J.F.; Zheng, Y.G.; Qin, Y.J.; Gao, W.W.; Li, G.Y. Cavitation erosion behavior and mechanism of HVOF sprayed WC-10Co-4Cr coating in 3.5 wt.% NaCl solution. *Trans. Indian Inst. Met.* **2015**, *68*, 151–159. [CrossRef]

27. Zhao, X.Q.; Zhou, H.D.; Chen, J.M. Comparative study of the friction and wear behavior of plasma sprayed conventional and nanostructured WC-12%Co coatings on stainless steel. *Mater. Sci. Eng. A* **2006**, *431*, 290–297. [CrossRef]

28. Wang, H.T.; Chen, X.; Bai, X.B.; Ji, G.C.; Dong, Z.X.; Yi, D.L. Microstructure and properties of cold sprayed multimodal WC-17Co deposits. *Int. J. Refract. Met. Hard Mater.* **2014**, *45*, 196–203. [CrossRef]

29. Ding, X.; Cheng, X.D.; Yuan, C.Q.; Shi, J.; Ding, Z.X. Structure of micro-nano WC-10Co4Cr coating and cavitation erosion resistance in NaCl solution. *Chin. J. Mech. Eng. Engl. Ed.* **2017**, *30*, 1239–1247. [CrossRef]

30. Ding, X.; Cheng, X.D.; Yu, X.; Li, C.; Yuan, C.Q.; Ding, Z.X. Structure and cavitation erosion behavior of HVOF sprayed multi-dimensional WC-10Co4Cr coating. *Trans. Nonferrous Met. Soc. China* **2018**, *28*, 487–494. [CrossRef]

31. Al-Mutairi, S.; Hashmi, M.S.J.; Yilbas, B.S.; Stokes, J. Microstructural characterization of HVOF/plasma thermal spray of micro/nano WC-12%Co powders. *Surf. Coat. Technol.* **2015**, *264*, 175–186. [CrossRef]

32. Sahraoui, T.; Guessasma, S.; Jeridane, M.L.; Hadji, M. HVOF sprayed WC-Co coatings: Microstructure, mechanical properties and friction moment prediction. *Mater. Des.* **2010**, *31*, 1431–1437. [CrossRef]

33. *GB/T 6383-2009 The Method of Vibration Cavitation Erosion Test*; Standards Administration of China: Beijing, China, 2009.

coatings

MDPI

Article

Hardening of HVOF-Sprayed Austenitic Stainless-Steel Coatings by Gas Nitriding

Thomas Lindner *, Pia Kutschmann, Martin Löbel and Thomas Lampke

Materials and Surface Engineering Group, Institute of Materials Science and Engineering, Chemnitz University of Technology, D-09107 Chemnitz, Germany; pia.kutschmann@mb.tu-chemnitz.de (P.K.); martin.loebel@mb.tu-chemnitz.de (M.L.); thomas.lampke@mb.tu-chemnitz.de (T.L.)
* Correspondence: th.lindner@mb.tu-chemnitz.de; Tel.: +49-371-531-38287

Received: 24 August 2018; Accepted: 27 September 2018; Published: 29 September 2018

Abstract: Austenitic stainless steel exhibits an excellent corrosion behavior. The relatively poor wear resistance can be improved by surface hardening, whereby thermochemical processes offer an economic option. The successful diffusion enrichment of bulk material requires a decomposition of the passive layer. A gas nitriding of high velocity oxygen fuel spraying (HVOF)-sprayed AISI 316L coatings without an additional activation step was studied with a variation of the process temperature depending on the heat-treatment state of the coating. A successful nitrogen enrichment was found in as-sprayed condition, whereas passivation prevents diffusion after solution heat treatment. The phase composition and microstructure formation were examined. The crystal structure and lattice parameters were determined using X-ray diffraction analysis. The identified phases were assigned to the different microstructural elements using the color etchant Beraha II. In as-sprayed condition, the phase formation in the coating is related to the process temperature. The formation of the S-phase with interstitial solvation of nitrogen is achieved by a process temperature of 420 °C. Precipitation occurs during the heat treatment at 520 °C. In both cases, a significant increase in wear resistance was found. The correlation of the thermochemical process parameters and the microstructural properties contributes to a better understanding of the requirements for the process combination of thermal spraying and diffusion.

Keywords: thermal spraying; thermochemical treatment; nitriding; hardening; S-phase; AISI 316L; stainless steel; high velocity oxygen fuel spraying (HVOF)

1. Introduction

Surface hardening of austenitic stainless steel in bulk material state involves an activation step degrading the passivation layer, which prevents surface diffusion. Depending on the temperature–time profile, the phase formation for carbon or nitrogen enrichment differs. Increasing process temperature and duration can cause the formation of precipitates that reduce the corrosion resistance due to chromium depletion. Low-temperature thermochemical treatment can permit a supersaturation of the crystal matrix by interstitial elements. Carbon and nitrogen can diffuse into the interstices of the face-centred cubic (FCC) lattice after dissolving the passivation layer [1–3]. These heat treatment processes are mainly used in the case of bulk material treatment [4]. Qualifying a process combination of coating technologies and additional surface hardening can be a choice for new applications. The separation of bulk and surface properties enables a multi-material design and a better adaption to local requirements. Furthermore, repair solutions are possible. In particular, thermal spraying offers the possibility to prevent a degradation of the metastable S-phase layer by limitation of heat input.

The fundamental feasibility of combining the processes of thermal spraying for deposition of austenitic stainless-steel coatings and thermochemical treatment has been shown in several studies [4–11]. However, the need of using the special process management of thermochemical treatment with the activation step designed for bulk material is not proven for thermally sprayed coatings. The reactions occurring while processing the austenitic feedstock material as well as the microstructure of thermally sprayed coatings raise doubts on the expediency of the parameter settings for thermochemical treatment.

Nestler et al. investigated the diffusion enrichment of high velocity oxygen fuel spraying (HVOF)-sprayed AISI 316L coatings in a gas nitriding process without an additional activation step. The formation of a compound layer by nitrogen diffusion from the surface into the coating was proven in a temperature range between 500 °C and 580 °C. The intended enrichment of the substrate material through the coating was not achieved. They assumed that a nitrogen transport through the coating structure is suppressed by the effect of low porosity and chemical composition [5]. On the other hand, Park et al. focussed on the plasma nitriding and nitro-carburizing process to improve the wear resistance of AISI 316L coatings by precipitation hardening [6]. After nitriding, the compound layer consists of CrN, Fe_3N and Fe_4N, but the diffusion depth is lower compared to the findings of Nestler et al. [5,6].

Wielage et al. first mentioned a successful S-phase formation for HVOF-sprayed AISI 316L coatings without chromium depletion by low temperature carburization with an additional surface-activation step [7]. Adachi et al. and Lindner et al. investigated further methods of low-temperature thermochemical processes. Adachi et al. carried out several studies on AISI 316L coatings produced by different thermal-spraying processes and a subsequent plasma process to enrich the coating surface with nitrogen, carbon, or both. They found that the diffusion-layer thickness is influenced by the temperature, whereas the coating condition and microstructure do not promote the interstitial diffusion [4,8–10]. Lindner et al. applied a gas nitro-carburizing process on HVOF-sprayed AISI 316L coatings and determined a certain porosity effect on diffusion depth [11]. Both noticed that the low-temperature processes enable the formation of the S-phase at regions near the surface area in thermally sprayed austenitic stainless-steel coatings with a diffusion depth slightly larger or comparable to bulk material. A lattice expansion of the austenitic phase was found by X-ray diffraction due to the interstitial solution of nitrogen and carbon atoms [7,11]. The described structural effects for thermally sprayed coatings are comparable to the results in bulk material state [12,13]. The dependency on the considered crystal planes follows the anisotropic elastic behavior of austenite, whereby enrichment of nitrogen causes a higher distortion than carbon. Both compound-layer formation by precipitation hardening and S-phase formation by interstitial solvation result in a significant improvement of the wear resistance of thermally sprayed stainless-steel coatings [4,5,7–11].

The present study focuses on the hardenability of HVOF-sprayed AISI 316L coatings by thermochemical treatment without an initial surface activation. A gas-nitriding process was performed using different temperature–time profiles aiming at precipitation and solution hardening. The diffusion behavior and phase formation were investigated with regard to the coating's structure. The wear behavior was determined for the untreated and the gas-nitrided states. A solution heat treatment of coatings was performed to reduce deviations in the element concentration caused by thermal-spray processing. The proven change of diffusion behavior permits alternative parameter settings for the thermochemical treatment of thermally sprayed coatings. The results contribute to a better understanding of material behavior in conjunction with the S-phase formation of austenitic stainless-steel coatings.

2. Materials and Methods

As coating material, gas-atomized AISI 316L powder was used in a fraction of −53+20 µm. Substrates were prepared out of the same kind of stainless steel using bar material with a diameter of 40 mm in cuts of 8 mm. After grid blasting the front surface using EK-F 24 (3 bar, 20 mm distance, 70° angle), the samples were cleaned for 5 min in an ultrasonic ethanol bath. The thermally sprayed coatings were produced by a HVOF K2 system (GTV, Luckenbach, Germany) with the parameters shown in Table 1. With an average coating thickness of 297 μm after 12 passes, the mean growth rate per pass is approximately 25 μm.

Table 1. The parameters of the HVOF thermal-spraying process with a GTV-K2 system.

Kerosene Flow Rate (L/h)	Oxygen Flow Rate (L/min)	λ	Combustion Chamber Pressure (bar)	Spraying Distance (mm)	Nozzle (mm)	Transverse Speed (m/s)	Offset (mm)	Powder Feed Rate (g/min)	Gas Feed (Argon) (L/min)	Wiper
24	900	1.1	7.1	350	150/14	1	5	70	2×8	NL

The average chemical composition of the main alloying elements was analyzed by X-ray fluorescence (XRF) utilizing a FISCHERSCOPE X-RAY XAN (Helmut Fischer, Sindelfingen, Germany) with 30 kW and 1 mm collimator lens for substrate, feedstock and coating material. The evaluation software Fischer XAN-WinFTM 6.33 (Helmut Fischer, Sindelfingen, Germany) was used. Five positions were measured for each sample. Furthermore, a quantification of the oxygen content was carried out using carrier-gas hot extraction (CGHE) TC600 (Leco, St. Joseph, MI, USA).

While it can be expected that the passivation layer prevents a diffusion enrichment of the austenitic stainless-steel bulk sample, additionally a ferritic steel grad AISI 1015 was chosen as further reference. The surface of the coated samples and the bulk material were polished up to mesh 1000. The final coating thickness is approximately 270 μm. A solution heat treatment at 1100 °C for 4 h was performed in a Torvac 12 Mark IV furnace under vacuum (10^{-4} mbar). The bulk references and the coatings in heat-treated and as-sprayed condition were gas nitrided in an industrial vacuum chamber retort WMU heat-treatment furnace using the parameters shown in Table 2. The different treatment temperatures were adapted to the different hardening techniques.

Table 2. Gas-nitriding parameters.

Process Gas	Volume Flow (L/h)	Temperature (°C)	Duration (h)
NH_3	1000	520; 420	36; 30

Metallographic cross-sections were prepared according to standard metallographic procedures. After hot mounting the section parts in conductive resin, grinding, and polishing, the samples were etched using Beraha II color etchant to visualize the different microstructural constituents. The distinguished color contrast gives evidence of diffusion enrichment as well as their presence and distribution. The metallographic investigations were performed using an optical microscope GX51 (Olympus, Shinjuku, Japan) equipped with a SC50 camera (Olympus, Shinjuku, Japan).

The microhardness was measured on the cross-section using a FISCHERSCOPE HM2000 XYm (Helmut Fischer, Sindelfingen, Germany) with 0.01 kp. The diffraction studies were performed by X-ray diffraction (XRD) using a D8 DISCOVER diffractometer (Bruker AXS, Billerica, MA, USA). Co-Kα radiation was used with a tube voltage of 40 kV, a tube current of 40 mA, polycap optics for beam shaping, a 1 mm pinhole collimator, and a 1D Lynxeye XE detector (Bruker, MA, USA). The diffraction diagrams were measured in the diffraction angle (2θ) range from 20° to 129° with a step size of 0.01°, and 1.5 s/step, which corresponds to 288 s/step due to the use of the 1D detector. The powder diffraction file (PDF) database 2014 was applied for phase identification and the determination of the lattice parameters.

The tribological properties of the coatings were investigated with the reciprocating ball-on-plane test leant on ASTM G 133 [14] as dry couple and the ball-on-disk test leant on ASTM G 99 [15] as a dry sliding system, both with a spherical counterbodies. The test parameters as well as the counterbody material and diameter are shown in Table 3. The wear volumes and depths were determined with the

3D profilometer MikroCAD (LMI, Teltow, Germany). The wear coefficient K was calculated according to Equation (1), with the wear volume V, the sliding distance l and the normal load F.

$$K = \frac{V}{F \cdot l} \qquad (1)$$

Table 3. Testing parameters of wear tests.

Reciprocating Ball-on-Plane Test		Ball-on-Disk Test	
Normal load	26 N	Normal load	20 N
Frequency	40 Hz	Radius	5 mm
Time	900 s	Speed	96 RPM
Amplitude	0.5 mm	Cycles	15916
Ø Al_2O_3	10 mm	Ø Al_2O_3	6 mm

3. Results

3.1. Chemical Composition

The XRF measurements confirm a good agreement between the average chemical composition of the feedstock material and the coatings. Due to the coating process, minor reductions of chromium and molybdenum contents occur. Slight deviations can be found in comparison to the bulk sample. A lower nickel content was observed along with a simultaneous increase in iron concentration. The CGHE quantifies an increase in oxygen content for the HVOF coating. This is related to the atmospheric condition during processing the feedstock, as shown in Table 4.

Table 4. Average chemical composition of the AISI 316L samples in different conditions, measured by XRF and CGHE (main alloy elements in relative wt.%, oxygen in absolute wt.%).

Sample	Feedstock	HVOF Coating	Substrate
Fe	66.2	67	70
Cr	16.7	16.5	16.8
Ni	12.8	12.7	10.1
Mo	2.8	2.3	1.9
Mn	1.5	1.5	1.2
O_2	0.1	0.65	–

3.2. Microstructure Analyses

The microstructure of the coating is affected by the solution treatment step. Figure 1 shows the difference in microstructure formation in cross-section. In as-sprayed condition, the single spray particles can be clearly distinguished by the contrast of cluster boundaries including some porosity. The contrast indicates possible oxidation of alloy components at the surface of the single particles. By solution heat treatment, segregations were dissolved, whereas porosity remains.

A successful diffusion enrichment without an initial activation step was found for the AISI 316L HVOF coating in as-sprayed condition. Figure 2 shows the cross-section of the coatings in dependence of the used gas-nitriding parameters. The domains in the microstructure can be distinguished depending on the process temperature using Beraha II color etchant. Within the coating, three phase areas were determined, proving a heterogeneous microstructure. The S-phase appears white, while a brown or blue color indicates the initial austenitic phase. In accordance with the results of Nestler et al., a compound layer is formed on top of the coating at 520 °C [5]. This layer possesses an intense reaction with the color etchant causing a deep contrast. Underneath, a thin, white layer can be detected, which can be assigned to the S-phase. Due to the different etching technique, Nestler et al. do not prove this in their research. Several isolated white particles with a high surface distance indicate a permeability of the coating. Moreover, the compound-layer thickness above 100 μm proves the influence of the coating porosity, as an AISI 1015 bulk reference shows only 20 μm. Furthermore, a compound-layer thickness of 40 μm and 44 μm was described by Rajendran et al. and Subbiah et al. after gas nitriding of AISI 316LN bulk material by using similar process parameters [16,17].

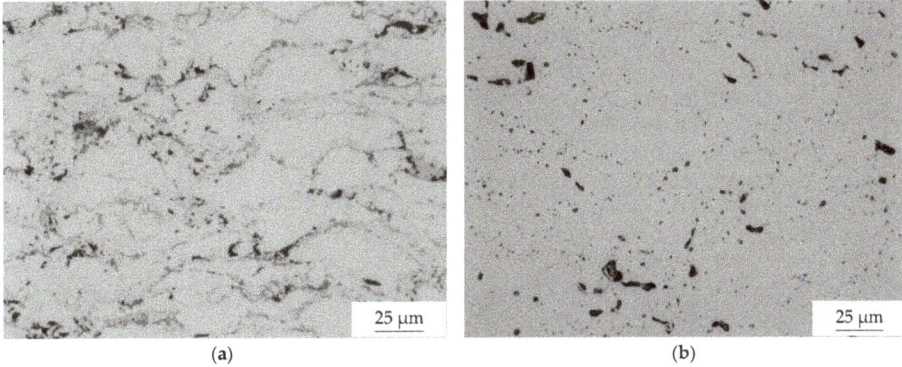

(a)   (b)

Figure 1. Optical-microscopic images of the AISI 316L HVOF coating in cross-section (a) as-sprayed condition and after (b) solution heat treatment.

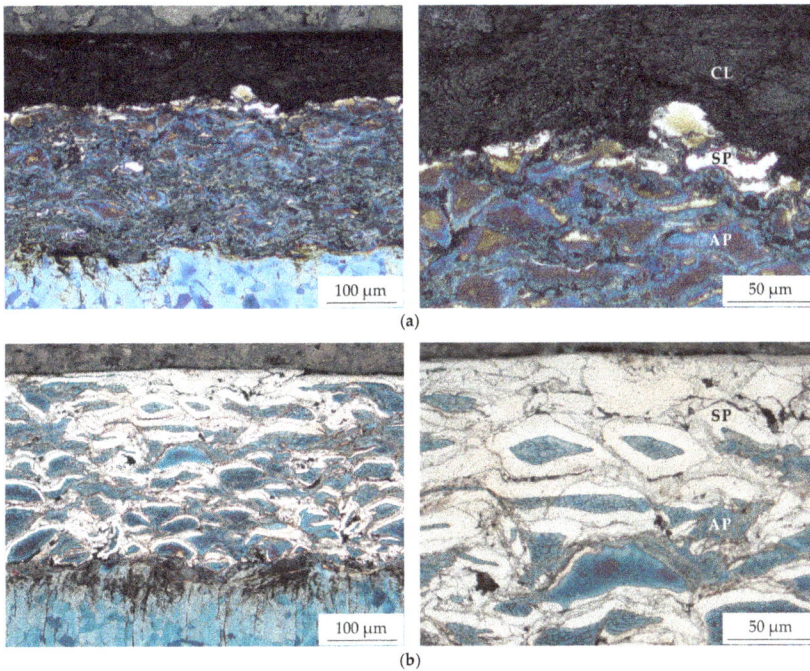

(a)

(b)

Figure 2. Optical-microscopic images of a Beraha II-etched AISI 316L HVOF coating in as-sprayed conditions after gas nitriding with different parameters (a) gas nitrided, 520 °C, 36 h and (b) gas nitrided, 420 °C, 30 h in overall view and detail with phase declaration (CL: compound layer; SP: S-phase; AP: austenitic phase).

The reduction of the temperature level and duration of the gas-nitriding process causes a change in phase formation. Precipitates can be avoided, whereas S-phase is formed at the surface of the spray particles, confirming results of gas nitro-carburization [11]. The S-phase can be found between the coating's surface and the coating–substrate interface. The diffusion depth within the single spray particles slightly decreases with increasing distance from the surface, which can be explained by a limitation of exchangeability. Compound-layer formation causes a volume increase that reduces the permeability.

The passive layer was repeatedly mentioned to be an effective diffusion barrier. The results of gas nitriding at 420 °C (Figure 3) prove that no diffusion enrichment can be observed for the bulk material without an activation step. Furthermore, the solution-heat-treated coatings show equal material behavior. No deviation in etchant color occurs, indicating a homogeneous phase distribution. This suggests a homogeneous passivation layer with diffusion protection of the surface. It can be assumed that deviations in element concentrations as well as a heterogeneous oxygen enrichment within the coating are the main reason for the change in diffusion behavior.

(a)   (b)

Figure 3. Optical-microscopic images of a Beraha II-etched bulk sample and the solution-heat-treated AISI 316L HVOF coating after gas nitriding 420 °C/30 h in cross-section (TS: thermally sprayed coating; SM: substrate material) (**a**) bulk sample and (**b**) solution heat treated HVOF coating.

An increase in microhardness is caused by the phase transformation. The unaffected austenitic phase shows an average value of 334 HV 0.01, whereas the highest value was detected for the S-phase with a maximum at 1120 HV 0.01. No differences were observed depending on the distance to the surface. In accordance with this, the compound layer shows no gradient, whereas a slightly reduced microhardness was found. Table 5 summarizes the microhardness values of the different phases with their standard deviations.

Table 5. Microhardness of the different phases in the AISI 316L HVOF coating.

Phase	Austenite	S-phase	Compound Layer
HV 0.01	334 ± 15	958 ± 86	861 ± 102

Depending on process parameters of the thermochemical treatment, a change in phase formation was observed by XRD studies, as shown in Figure 4. In comparison to the untreated sample, a shift of the characteristic peaks of the austenitic phase to lower angles was determined for the sample thermochemical treated at 420 °C. This can be explained by the increase of lattice spacing due to the interstitial inclusion of nitrogen atoms in the mixed crystal. Due to the different Young's moduli, the expansion of the lattice planes {111} and {200} differs. The lattice parameter of the austenitic phase extends from $a = 3.59$ Å to $a = 3.92$ Å for the planes {111}. Precipitates can be excluded for the thermochemical treatment at 420 °C, but nitride phases occur at 520 °C. The characteristic peaks can be assigned to iron and chromium nitride phases.

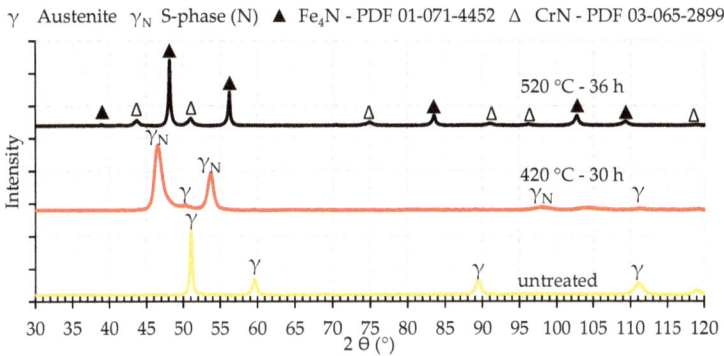

Figure 4. Diffraction diagrams of the gas nitrided and untreated AISI 316L HVOF coating in as-sprayed condition.

A significant increase in wear resistance of the coating was found for both gas-nitriding conditions, as shown in Figure 5. A reduction of the wear coefficient of more than 90% for sliding ball-on-disk test and approximately 75% for reciprocating ball-on-plane test was detected. As a consequence, the maximum values of wear depth were reduced in a similar manner.

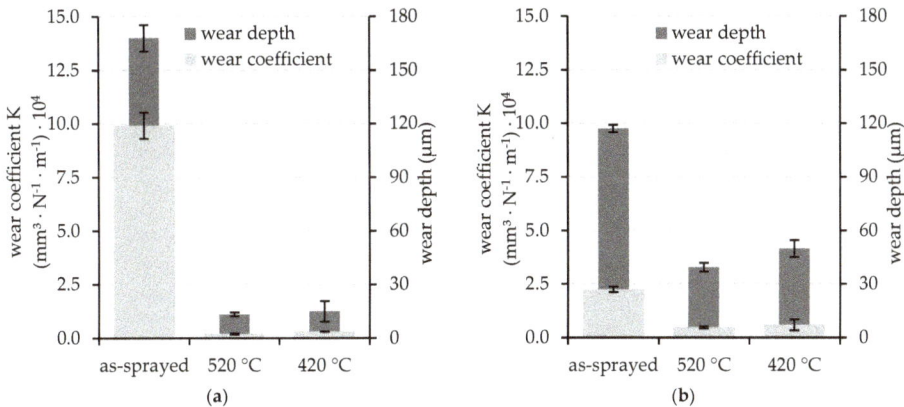

Figure 5. Wear values of the AISI 316L HVOF coating depending on the state of thermochemical treatment for different wear tests (**a**) ball-on-disk tests and (**b**) reciprocating ball-on-plane tests.

4. Summary

The results of the treatment of the coatings in as-sprayed condition prove that no additional activation step is required for successful diffusion enrichment by gas nitriding. This derives from the different material behaviors of bulk material and the coating. Phase formation depends on the temperature level and the duration time. Precipitation occurs under elevated temperatures, whereas the solid solution is supersaturated at a lower process temperature. A homogeneous precipitation layer grows from the coating's surface, whereas the S-phase is formed on the surface of the individual spray particles distributed between the surface and substrate. The coating's permeability allows for the penetration of gaseous enrichment media causing structural diffusion. A solution heat treatment of the coating reduces heterogeneity and a protective passive layer is formed. Solution and precipitation

Coatings **2018**, *8*, 348

hardening significantly improve the wear resistance. However, the differences between the various hardening mechanisms are minor, whereby S-phase formation prevents chromium depletion.

Author Contributions: T.L. (Thomas Lindner) and P.K. conceived and designed the experiments. T.L. (Thomas Lindner), P.K. and M.L. performed the experiments, analysed the data and wrote the paper. T.L. (Thomas Lampke) directed the research and contributed to the discussion and interpretation of the results.

Funding: This research received no external funding.

Acknowledgments: The authors thank Thomas Mehner for conducting the XRD measurements and Gunar Röllig for support in wear testing.

Conflicts of Interest: The authors declare no conflict of interest.

References

1. Somers, M.A.J.; Christiansen, T.L. Low temperature surface hardening of stainless steel. In *Thermochemical Surface Engineering of Steels*; Mittemeijer, E.J., Somers, M.A.J., Eds.; Elsevier Ltd.: Amsterdam, The Netherlands, 2015; pp. 557–579.
2. Zhao, C.; Li, C.X.; Dong, H.; Bell, T. Low temperature plasma nitrocarburising of AISI 316 austenitic stainless steel. *Surf. Coat. Technol.* **2005**, *191*, 195–200.
3. Bell, T. Current status of supersaturated surface engineered S-phase materials. *Key Eng. Mater.* **2008**, *373*, 289–295. [CrossRef]
4. Adachi, S.; Ueda, N. Combined plasma carburizing and nitriding of sprayed AISI 316L coating for improved wear resistance. *Surf. Coat. Technol.* **2014**, *259*, 44–49. [CrossRef]
5. Nestler, M.C.; Spies, H.; Hermann, K. Production of duplex coatings by thermal spraying and nitriding. *Surf. Eng.* **1996**, *12*, 299–302. [CrossRef]
6. Park, G.; Bae, G.; Moon, K.; Lee, C. Effect of plasma nitriding and nitrocarburinzing on HVOF-sprayed stainless steel coatings. *J. Therm. Spray Technol.* **2013**, *22*, 1366–1373. [CrossRef]
7. Wielage, B.; Rupprecht, C.; Lindner, T.; Hunger, R. Surface modification of austenitic thermal spray coatings by low-temperature carburization. In Proceedings of the International Thermal Spray Conference & Exposition, Long Beach, CA, USA, 11–14 May 2015. DVS 276.
8. Adachi, S.; Ueda, N. Formation of S-phase layer on plasma sprayed AISI 316L stainless steel coating by plasma nitriding at low temperature. *Thin Solid Films* **2012**, *523*, 11–14. [CrossRef]
9. Adachi, S.; Ueda, N. Formation of expanded austenite on a cold-sprayed AISI 316L coating by low-temperature plasma nitriding. *J. Therm. Spray Technol.* **2015**, *24*, 1399–1407. [CrossRef]
10. Adachi, S.; Ueda, N. Surface hardness improvement of plasma-sprayed AISI 316L stainless steel coating by low-temperature plasma carburizing. *Adv. Powder Technol.* **2013**, *24*, 818–823. [CrossRef]
11. Lindner, T.; Mehner, T.; Lampke, T. Surface modification of austenitic thermal-spray coatings by low-temperature nitrocarburizing. *IOP Conf. Ser. Mater. Sci. Eng.* **2016**, *118*, 012008. [CrossRef]
12. Christiansen, T.L.; Hummelshøj, T.S.; Somers, M.A.J. Expanded austenite, crystallography and residual stress. *Surf. Eng.* **2010**, *26*, 242–247. [CrossRef]
13. Brink, B.K.; Ståhl, K.; Christiansen, T.L.; Oddershede, J. On the elusive crystal structure of expanded austenite. *Scr. Mater.* **2017**, *131*, 59–62. [CrossRef]
14. *ASTM G 133 Standard Test Method for Linearly Reciprocating Ball-on-Flat Sliding Wear*; ASTM International: West Conshohocken, PA, USA, 2016.
15. *ASTM G 99 Standard Test Method for Wear Testing with a Pin-on-Disk Apparatus*; ASTM International: West Conshohocken, PA, USA, 2016.
16. Rajendrana, P.; Devaraju, A. Experimental evaluation of mechanical and tribological behaviours of gas nitride treated AISI 316LN austenitic stainless steel. *Mater. Today Proc.* **2018**, *5*, 14333–14338. [CrossRef]
17. Subbiah, R.; Rajavel, R. Dry sliding wear behaviour analysis of nitrided 316LN grade austenitic stainless steels using gas nitriding process. *J. Theor. Appl. Inf. Technol.* **2010**, *19*, 98–101.

coatings

MDPI

Article

Investigating the Sanding Process of Medium-Density Fiberboard and Korean Pine for Material Removal and Surface Creation

Jian Zhang [1,2], Junhua Ying [1,2], Feng Cheng [1,2], Hongguang Liu [1,2], Bin Luo [1,2,*] and Li Li [1,2]

[1]   College of Materials Science and Technology, Beijing Forestry University, Beijing 100083, China; wujijianshuai2016@bjfu.edu.cn (J.Z.); yingjunhua66@163.com (J.Y.); chengfeng@bjfu.edu.cn (F.C.); bjfuliuhg@bjfu.edu.cn (H.L.); depwoodlili@bjfu.edu.cn (L.L.)

[2]   MOE Key Laboratory of Wooden Materials Science and Application, College of Materials Science and Technology, Beijing Forestry University, Beijing 100083, China

*   Correspondence: luobincl@bjfu.edu.cn

Received: 10 September 2018; Accepted: 21 November 2018; Published: 22 November 2018

Abstract: As an important fine machining method, sanding operation is widely used in most engineered materials. In wood sanding, high material removal rate and surface quality are expected. Clarifying the material deformation in the sanding process is the key to improving sanding efficiency. In this study, a single grit scratching method is used to investigate the material removal and surface creation of medium-density fiberboard (MDF) and Korean Pine (Pinus koraiensis Sieb.et Zucc). It is found that there are some differences in the material deformation during scratching Korean Pine and MDF, compared with grinding metals. A mechanism based on the anatomical cavities absorbing effect was proposed to account for the differences. This mechanism helps to explain why tiny, or even no, "pile-up" (like swelling ridges created by the ploughing effect) occurs during scratching Korean Pine, especially in longitudinal direction. MDF as a densified wood composite presented more pile-up and the variation of pile-up ratio was investigated. The porosity and wood grain direction exert great influence on material removal and surface creation in wood sanding. At the rubbing stage, a new method was developed to confirm the elastic spring back effect both in MDF and Korean Pine scratching. The results obtained and the approaches used in this paper could provide insights into the material removal and surface creation research of other wood species and wood composites to finally improve sanding efficiency and surface quality.

Keywords: wood sanding; material deformation; surface morphology; ploughing; elastic spring back

---

1. Introduction

Wood materials are widely used in floor and furniture manufacturing. As an important fine machining method, sanding of wood is usually the last operation of surface modification before gluing and painting, which aims to improve surface quality and ensure dimensional accuracy [1–3]. Looking through related literatures, there are not many systematic studies on material removal and surface creation in wood sanding operation. Stewart studied some surfacing defects and problems related to wood moisture content [4]. He proposed that the defects commonly associated with sanding operations are fuzzy or raised grain, crushed or burned surface, and moisture content exerts great influence on it. Grain raising like lifting of fibres in wood was deeply researched by Evans et al. [5]. They clarified the relationship between wood density and grain raising. In addition, Koehler observed that sanded wood surface was more susceptible to grain raising than planed wood surfaces [6].

Despite the studies on sanded surface morphology and defects, the relationship between the key variables (grit size, cutting depth, feed rate, sanding speed, and grain direction) and material removal rate and resultant surface roughness was investigated in abrasive belt sanding operations [7–9]. Sulaiman et al. studied the effect of sanding on surface roughness of rubberwood [3]. It was found that lower sanding grit size gave rougher surface and rougher surface was more wettable compared to smoother surface. The contact angle on the tangential surface was higher compared to radial surface. In Cool and Hernandez's research, the sanding of black spruce wood prior to coating application was optimized for feed speed and grit size [10]. It was found that coarser grit size and higher feed speeds contributed to increased surface roughness and improved wetting properties. However, Sun and Li observed that the glue bonding strength was not necessarily increased as the surface became smoother for solid wood [11].

For a long period of time, the classical grinding theory of metal is taken as a reference in wood sanding. Considering the grit-workpiece interaction, Hahn divided the material deformation into three phases which are rubbing, ploughing and cutting [12]. Rubbing occurs at the initial engagement of grit into workpiece, with only elastic and slightly plastic deformation of workpiece material. When the grit penetrates deeper into the workpiece, ploughing takes place and clearly visible groove traces with lateral bulging are formed. With the increasing cutting depth, the shearing stress at the ploughed material ahead of the grit increases and a chip is formed, which is called cutting. Analogously, Klocke presented a model of chip formation which consists of three phases [13]. In the first phase, elastic deformation takes place. As the grit penetrates deeper, plastic deformation occurs as well as elastic deformation. When the cutting depth reaches a critical value, the chip removal begins in the third phase, where elastic-plastic deformation and material removal occur superimposed.

For metal, alloy and ceramics, the grinding researches are mainly about material removal and surface integrity from a microcosmic view. Chen and Öpöz presented an investigation of grinding material removal mechanism using finite element (FE) method [14]. In this research, material removal mechanism of grinding, namely rubbing, ploughing and cutting, was discussed with the variation friction coefficient. Anderson et al. compared the spherical and conical single grit on 4340 steel workpieces cutting tests [15]. Meanwhile, a three-dimensional finite element (FE) model was created. It was pointed out that grit cutting edge shape influenced the ploughing and cutting actions significantly [16,17].

The structure of metals is typically considered homogeneous, with the polycrystalline grains equally distributed throughout the material [18]. However, wood materials are composed of cellular structures with other properties such as anisotropy, high porosity and inhomogeneity [19]. Therefore, wood materials vary a lot in material removal and surface creation compared with metals grinding. In this study, the primary aim is to specify both the material elastic and plastic deformation, and to clarify how the deformed material influences the surface morphology. The anatomical cavities absorbing effect was proposed in terms of scratching MDF and Korean Pine. And that will make it possible to explain the differences of material plastic deformation. Grit geometry and multi-grit interference will be further investigated using some other wood species. The ultimate goal is to obtain a complete understanding of the material removal and surface creation of wood materials in the sanding process, which can help develop better sanding technology to improve sanding efficiency and surface quality.

2. Materials and Methods

2.1. Experimental Setup

Figure 1 shows the single grit scratching test system, which was designed and manufactured to meet the needs of making successive scratches on the surface of the workpiece. The wheel was driven by a precise electrical spindle. The cutting forces were recorded by a 3-axis piezoelectric force sensor (KISTLER 3257A; Kistler Instrumente AG, Winterthur, Switzerland), and the other devices included a charge amplifier (KISTLER 5806; Kistler Instrumente AG, Winterthur, Switzerland) and a signal analyzer (NEC OMNIACE II RA2300; NEC Corporation, Tokyo, Japan), the detailed information on the experimental parameters and conditions is described in Table 1.

The workbench has a horizontal movement in X and Y directions and a vertical movement in Z direction. The workpiece and the force sensor were mounted on the workbench through a work fixture. In order to generate successive scratches with different grit cutting depths, a traverse scratching method was used in the study. At the end of workpiece, there was a height less than 1 mm between the upper surface of workbench and the lower surface of workpiece. And the angle of these two surfaces was less than 0.5°. Hence, the successive scratches were generated with increasing cutting depth when the workbench was in movement of Y direction.

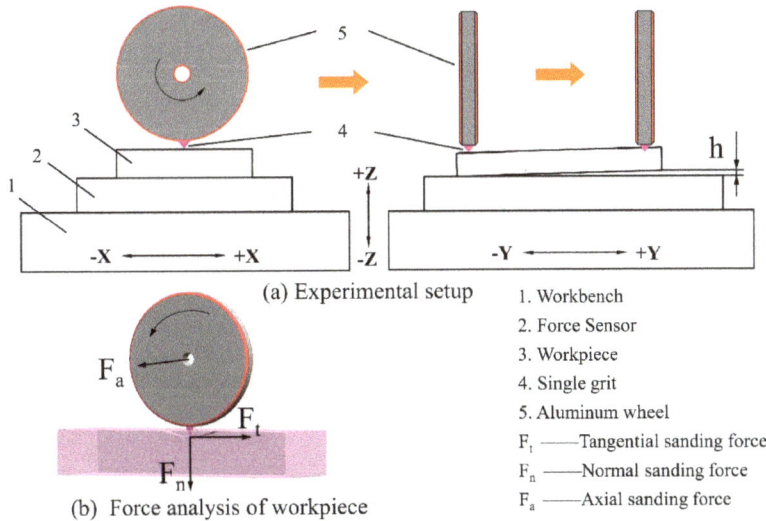

(a) Experimental setup

(b) Force analysis of workpiece

1. Workbench
2. Force Sensor
3. Workpiece
4. Single grit
5. Aluminum wheel
F_t ——Tangential sanding force
F_n ——Normal sanding force
F_a ——Axial sanding force

Figure 1. Schematic of the single grit scratching test system: (**a**) Experimental setup; (**b**) Force analysis of workpiece.

Table 1. Experimental parameters and conditions.

Item	Description
Feed velocity	$v_w = 9$ m/min
Cutting depth	0–0.5 mm
Electrical spindle	The power is 1.5 kW, the speed is 2100 r/min, with 0.01 mm accuracy
Grinding wheel	Made of aluminum, 0.2 kg in weight, the diameter is 100 mm, with 0.012 coaxial accuracy
3D force sensor	Sensitivity: $F_X \approx 7.5$ pC/N, $F_Y \approx -7.5$ pC/N, $F_Z \approx -3.7$ pC/N
Charge amplifier	Measuring range is from ±100 pC to $\pm1,000,000$ pC

2.2. Single Grit and Workpiece Preparation

The shape of actual corundum grit used in abrasive belt manufacturing is mostly three-pyramid or four-pyramid [20,21]. In this study, a three-pyramid grit was selected, as shown in Figure 2, to create scratches on workpiece surfaces. The detailed information of that grit is listed in Table 2. The grit was strongly glued onto the circumferential surface of the aluminum wheel by using super glue (ERGO 5800; Kisling AG, Wetzikon, Switzerland). In the process of all tests, the same grit was used without dropping off the wheel.

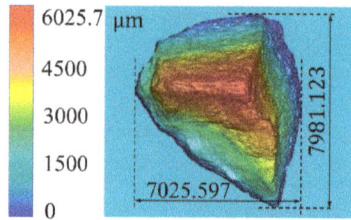

Figure 2. 3D morphology of the grit.

Table 2. The detailed information on the grit.

No.	Item	Results
1	Mesh	3
2	Al_2O_3	95.0%
3	SiO_2	1.0%
4	Fe	0.2%
5	Ti	2.0%
6	Density	3.6 g/cm^3
7	Mohs' hardness	$\geq$8.5

The materials selected for the test were medium-density fiberboard (MDF) and Korean Pine (Pinus koraiensis Sieb.et Zucc). The MDF was mainly made from Pinus branches using phenolic resin and the Korean Pine was sapwood about 30 years of age. Totally, 12 workpieces (4 for MDF, 8 for Korean Pine) were cut into the size of 130 mm (long) × 30 mm (wide) × 30 mm (thick). Densities of the workpieces were 0.78 ± 0.02 g/cm^3 for MDF and 0.33 ± 0.04 g/cm^3 for Korean Pine. All Korean Pine workpieces were dried in the electrical blast drying box (GZX-9070MBE; Shanghai Boxun Medical Biological Instrument Corp., Shanghai, China) before the experiments started to ensure final moisture content of 8% ± 2%. To precisely examine the microscopic morphology of scratches, the workpieces were levelled with an 80 grit aluminum oxide abrasive belt and then finish-sanded with 100 and 150 grit aluminum oxide abrasive belts using a wide belt-sander (SANDTEQ W-200, HOMAG, Schopfloch, Germany). An area measuring 24 × 18 mm^2 was scanned by 3D Profiler (KEYENCE VR-3200; KEYENCE, Osaka, Japan) and 4 different areas in each workpiece were scanned to obtain the surface roughness R_a of 2.4 ± 0.3 µm and 6.5 ± 0.7 µm for MDF and Korean Pine respectively.

2.3. The Scratch Direction aAnd Scratch Profile Measurement

MDF is generally made from different types of wood-based furnish such as pulp chips, planer shavings, plywood trims and sawdust. High mechanical strength and good water resistance are produced after the urea-formaldehyde (UF) or phenol-formaldehyde (PF) application and hot pressing [22]. The irregular spreading fibres and adhesive make it nearly isotropic. Korean Pine, as a kind of softwood, consists of axial tracheid cells, ray parenchyma and resin canals, with tracheids as the main component, making 90%–95% of the wood volume. These component cells vary widely in size [23,24]. In comparison with MDF, the characteristics of Korean Pine vary from each direction mainly because of the biological cell configuration and anatomy. Since material anisotropy exerts an important effect on the surface property, this study chose two directions (λ refers to the angle between the grit movement direction and wood grain direction) in scratching Korean Pine, as illustrated in Figure 3.

The scratch profiles on the workpiece were measured by 3D Profiler (KEYENCE VR-3200; KEYENCE, Osaka, Japan), with the measuring resolution of 0.5 µm on the Z axis. Each scratch was scanned after successive scratches were created on a workpiece surface as soon as possible to avoid the effect of air humidity variation. Figure 4a shows a typical 3D morphology of scratches, from which the characteristics of each individual scratch can be clearly seen. To further investigate the surface creation of MDF and Korean Pine, 2D cross-sectional profiles were extracted from the deepest point of scratches, as plotted in Figure 4b, where depth of cut, groove area, pile-up area (like swelling ridges created by the ploughing effect), and other related parameters can be measured by using analyzing software (VR Series version 2.2.0.89). Depth of cut means the height from the deepest point to baseline. Owing to the original biological surface roughness, the deepest point was generally in the middle of a scratch but sometimes can diverge from the middle zone. Pile-up ratio is the ratio of pile-up area to groove area of each scratch, which is widely used to measure the proportion of ploughing.

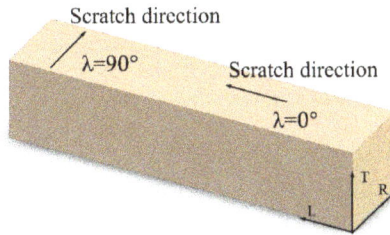

Figure 3. Schematic diagram of scratch direction and wood grain direction.

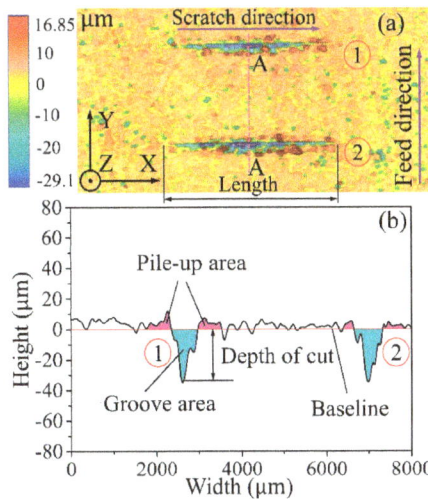

Figure 4. The measurement of scratches cut by single grit: (**a**) morphology of scratched workpiece surface; (**b**) cross-sectional profile of scratches.

2.4. The Grey Relational Grade Analysis (GRGA)

Since the characteristics of wood materials are complicated and can be affected by moisture content, grain direction, and anatomical structures, grey relational grade analysis (GRGA) was used in this study to figure out the specific influencing weight of depth of cut, groove area, pile-up area, and normal sanding force on pile-up ratio. The grey system theory was proposed by Deng [25]. Grey relational analysis (GRA) is broadly applied in evaluating or judging the performance of a complex project with insufficient information. The grey comprehensive relational grade reflects the similarity between polygonal lines X_0 and X_i geometrically, and it also reflects the proximity of the change rate of X_0 and X_i from the start point. Therefore, it is a good measure to indicate if there is a close relationship between the two sequences.

The processes of grey comprehensive relational grade analysis are as follows:

- The formation of the refence sequence and the compared sequence.

The reference sequence is marked as $X_0 = (x_0(1)x_0(2), \cdots, x_0(n))$, and the compared sequence is marked as $X_i = (x_i(1)x_i(2), \cdots, x_i(n))$, where $i = 1, 2, \cdots, m$ and the $x_i(n)$ is the n-th element of X_i.

- The calculation of the grey absolute relational grade.

Let $X_0^0 = \left(x_0^0(1)x_0^0(2), \cdots x_i^0(n)\right)$ and $X_i^0 = \left(x_i^0(1)x_i^0(2) \cdots x_i^0(n)\right)$ are the initialization of X_0 and X_i respectively. Mark:

$$|S_i| = \left|\sum_{k=2}^{n-1} x_i^0(k) + 1/2x_i^0(n)\right| \tag{1}$$

$$|S_i - S_0| = \left|\sum_{k=2}^{n-1} \left(x_k^0(k) - x_0^0(k) + 1/2x_i^0(n) - x_0^0(n)\right)\right| \tag{2}$$

where $i = 1, 2, \cdots, m; k = 1, 2, \cdots n$. Then, the grey absolute relational grade can be calculated by Equation (3):

$$\varepsilon_{0i} = \frac{1 + |S_0| + |S_i|}{1 + |S_0| + |S_i| + |S_i - S_0|} \tag{3}$$

where ε_{0i} is only related to the geometrical shape of X_0 and X_i. The bigger ε_{0i} value means that the geometrical shapes of X_0 and X_i in the line chart are more similar.

- The calculation of the grey relative relational grade.

Let $X'_i(k) = \frac{x_i(k)}{x_i(1)}$, then

$$|S'_i| = \left|\sum_{k=2}^{n-1} X'^0_i(k) + 1/2x'^0_i(n)\right| \tag{4}$$

where $i = 1, 2, \cdots, m; k = 1, 2, \cdots, n$. Then, the grey relative relational grade can be calculated by Equation (5):

$$R_{0i} = \frac{1 + |S'_0| + |S'_i|}{1 + |S'_0| + |S'_i| + |S'_i - S'_0|} \tag{5}$$

The grey relative relational grade characterizes the change rate of the two sequences X_0 and X_i from the start point. The bigger R_{0i} value means the more proximate the two change rates are.

- The calculation of grey comprehensive relational grade.

$$\rho_{0i} = \theta\varepsilon_{0i} + (1 - \theta)R_{0i} \tag{6}$$

where $\theta \in [0, 1]$. Generally, $\theta = 0.5$ is preferable. If more concern is taken into the proximity of broken lines geometry, a bigger θ can be selected. If more concern is taken into the changing rate, a smaller θ can be chosen.

3. Results and Discussion

3.1. The Plastic Material Deformation of Scratch Cut by Single Grit

Figure 5 shows the cross-sectional profiles of the scratched surface and original surface of MDF and Korean Pine workpiece. It is found that the "pile-up" was more obvious in scratching MDF. When scratching Korean Pine ($\lambda = 0°$), there was minute even zero, pile-up occurred on two sides of the groove. Moreover, it is noted that there was a significant difference in the shape of grooves. The groove shape of MDF scratches presented serrated edges mostly. However, the groove shape of Korean Pine ($\lambda = 0°$) scratches was relatively smooth.

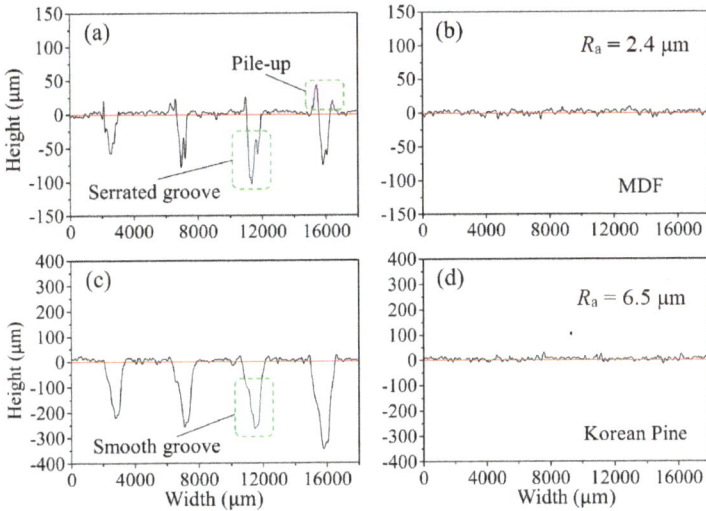

Figure 5. The cross-sectional profiles of scratched surface and original surface: (**a**) the scratched surface of MDF; (**b**) the original surface of MDF; (**c**) the scratched surface of Korean Pine ($\lambda = 0°$); (**d**) the original surface of Korean Pine.

Since wood is an anisotropic and porous material, and its pore structure affects its behavior more than any other characteristics [23,26], wood grain direction and unevenly distributed porosity (the amount of void volume) were considered to analyze the difference on plastic material deformation between scratching MDF and Korean Pine ($\lambda = 0°$). When scratching Korean Pine ($\lambda = 0°$), the scratch direction is parallel to the wood grain direction. Hence, with the grit movement along the grain direction, the anatomical cavities (axial tracheid cells, resin canals and ray parenchyma cells) are compressed in a direction perpendicular to wood grain direction. The elastic deformation will recover due to the elastic spring back effect after the grit passes. However, the material plastic deformation around the cutting-edge is mostly absorbed by the compressed cavities, which causes tiny even no pile-up, as illustrated in Figure 6a. Knowledge of the pore structure is directly related to the density, and wood density values are significantly low with the highest amount of porosity and the lowest amount of cell wall material [26,27]. In the process of making MDF, the natural structure of wood fibres is destroyed and many shorter fibres are mixed with adhesives in irregular distribution. In addition, the average density of MDF is about twice greater than Korean Pine, which means the porosity of Korea Pine is greater than that of MDF. Therefore, the capability of absorbing deformation for Korean Pine is better than MDF, which partially leads to more pile-up on bilateral sides of the groove in the case of scratching MDF, as shown in Figure 6b. The difference of groove shape may also be attributed to the absorbing effect.

The tiny "pile-up" occurred at bilateral sides of scratch (Korean Pine, $\lambda = 0°$) can be partly explained by the anatomical cavities absorbing effect. In addition, "fibres" are thin walled and are easily perforated and shredded during sanding, resulting in the formation of cell walls that are loosely bonded to the underlying wood surface, which was called "grain raising" especially in low-density wood species [5]. Figure 7a,b shows the microscopic images presented by Ultra-Depth 3D Microscope (KEYENCE VHX-6000, KEYENCE, Osaka, Japan) of scratched surface of MDF and Korean Pine ($\lambda = 0°$) workpiece, from which it can be inferred that the tiny "pile up" of Korean Pine scratched surface ($\lambda = 0°$) is more like "grain raising". Figure 7c shows the diagram of force analysis of grit-workpiece interaction, along the arch trajectory path, grit moves with negative rake angle until it reaches the deepest point and the cutting-edge of actual grit has a certain curvature radius. So, the sub-surface layer is compressed and densified mostly. When the plastic deformation exceeds the maximum capability of anatomical cavities absorption, there might be pile-up caused by the ploughing effect, which likes swelling ridges above the scratched surface. And, the shredded fibre structure accumulated in front of the rake face is removed if the F_n (the horizontal component of F) is big enough at the grit exit stage.

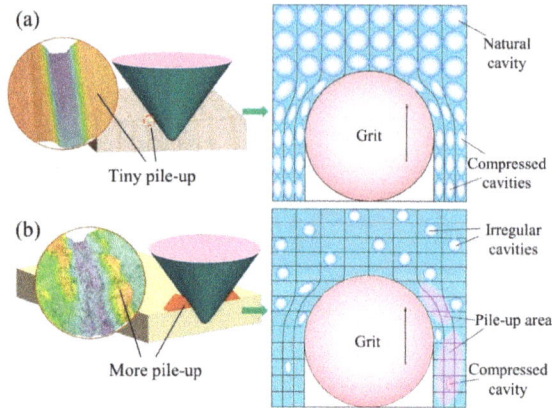

Figure 6. Schematic of the anatomical cavities absorbing effect (**a**) when scratching Korean Pine ($\lambda = 0°$) and (**b**) when scratching MDF.

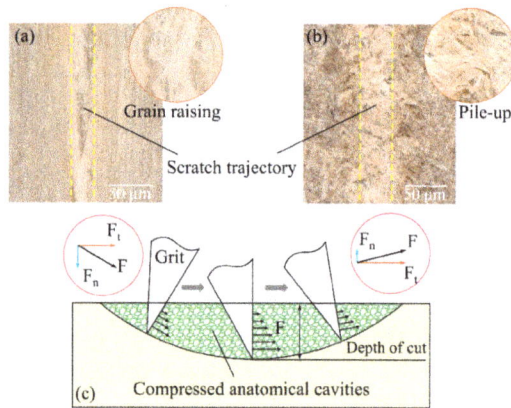

Figure 7. Analysis of plastic material deformation and chip formation when scratching Korean Pine ($\lambda = 0°$) and MDF: (**a**) Ultra-Depth 3D Microscopic images of Korean Pine scratched surface; (**b**) Ultra-Depth 3D Microscopic images of MDF scratched surface; (**c**) the diagram of force analysis of workpiece within the arch trajectory path.

As reported previously, the roughness average (R_a) was significantly higher in the direction perpendicular to the movement of the abrasive grains than along the grain [28]. Miao and Li found that the surface of Manchurian ash sanded in longitudinal direction seems to be the smoothest, while the surface of Birch sanded in transverse direction appears to have more wood wools [29]. From the surface morphology and cross-sectional profiles of scratching Korean Pine ($\lambda = 90°$), it is observed that groups of wood grain were not cleanly severed by scratching and stood up above bilateral sides of scratch (Figure 8a). In addition, the cross-sectional profiles of cracked wood materials were highly irregular, as the pink-colored area shown in Figure 8b. At the initial stage, the cracked wood grain was not obvious with low cutting depth. As the grit moved into the workpiece, more materials cracking occurred, especially around the deepest point of the scratch. Given this, the cracked grain occurred when scratching Korean Pine ($\lambda = 90°$) is more like a great amount of fuzziness rather than plastic pile-up [4]. It indicates that sanding direction exerts great influence on the surface morphology compared with the situation when $\lambda = 0°$.

To analyze the material removal and surface creation of scratching Korean Pine ($\lambda = 90°$), it is necessary to figure out the form of grit-workpiece interaction. In the case of $\lambda = 90°$, the rake face of the grit is strongly inclined to the workpiece surface when cutting depth is small at the initial stage [16]. At this time, the interaction between the grit cutting-edge and wood surface is point-contact or line-contact, and the contact length (W) is about zero (Figure 8c), where wood grain might be cut off but the fractured grain is hard to be removed in chips. Thus, the workpiece material is mostly crushed to two sides of the scratch. Cell crushing is caused by normal forces that outweighed the ultimate rupture strength of wood tissues [30]. The layer of superficial damaged tissues prevents the penetration of coatings or glues into the wood [31]. Moreover, crushed tissues at the surface and sub-surface might behave as a mechanical weak boundary layer in coating interfaces or gluelines [32,33]. As the grit moves forward, the interaction converts into face-contact, and the W becomes larger. In this case, wood grain can be cut off by side cutting-edges and removed in chips flow in front of the rake face when the shearing stress reaches the fracture strength. The fibre structure was transversely severed by side cutting-edges and caused much fuzzy grain standing up beside the scratch, which might account for the hugely irregular height profile (Figure 8b).

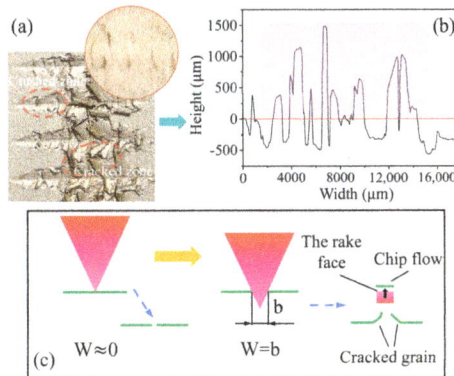

Figure 8. Analysis of plastic material deformation and chip formation when scratching Korean Pine ($\lambda = 90°$): (**a**) 3D morphology of Korean Pine scratched surface; (**b**) cross-sectional profiles of scratched surface; (**c**) the interaction form between grit cutting-edges and wood grain.

To further investigate the plastic material deformation, it was assumed that both the grain raising and cracked grain at the bilateral sides of scratches were considered as plastic pile-up. Then the effects of groove area and depth of cut on pile-up ratio when scratching MDF and Korean Pine are shown in Figure 9. When looking at the situation of scratching Korean Pine ($\lambda = 0°$), the pile-up ratio of both groove area and depth of cut were concentrated in a pink-colored belt with a range from 0 to 0.05. Through the case of scratching Korean Pine ($\lambda = 90°$), the pile-up ratio of both groove area and depth of cut were concentrated in a pink-colored belt with a range from 0 to 0.1, and it can be noted that the maximum pile-up ratio reached nearly 0.6. In these two circumstances, the ratios varied within a small range and showed no obvious or underlying trend with the variation of groove area and depth of cut. Actually, pile-up ratio is gradually decreasing while scratch depth is increasing with sharp cutting-edge grit [17]. Therefore, the assumption was not correct and the mechanism proposed above was confirmed from one aspect. When scratching MDF, the pile-up ratio presented to be scattered highly. And the polynomial fitting shows a decreasing tendency of pile-up ratio when the depth of cut and groove area increase. That means a greater proportion of cutting occurs with the increase of grit penetration depth.

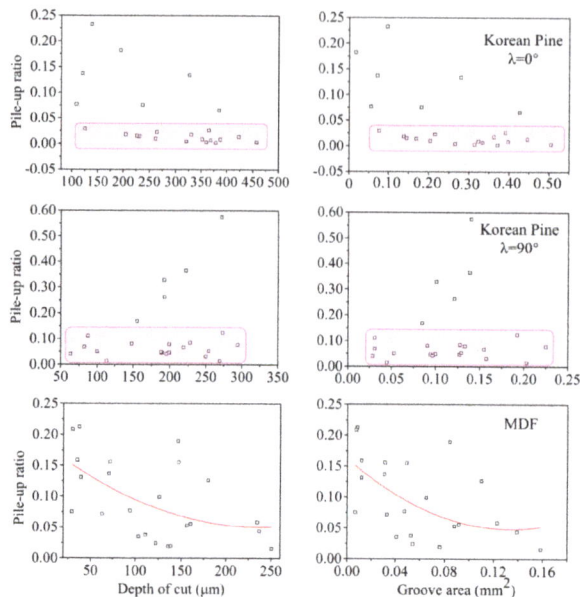

Figure 9. The effects of depth of cut and groove area on pile-up ratio.

To figure out the specific influencing weight of depth of cut, groove area, pile-up area and normal sanding force on pile-up ratio when scratching MDF, Grey relational grade analysis (GRGA) was used. In Equation (6), 0.2 was set as the value of θ for the dimensions of experimental statistics are different and the statistics vary a lot. Then the pile-up ratio was set as the reference sequence X_0; the depth of cut, groove area, pile-up area and normal sanding force were set as the compared sequence X_1, X_2, X_3 and X_4 respectively. According to the results listed in Table 3, the four grey comprehensive relational grades are greater than 0.5, which indicates that they are closely related to pile-up ratio. Specifically, the normal sanding force exerts the greatest influence on the pile-up ratio. This is because normal sanding force directly reflects the density (porosity), the other indices, however, are the indirect results after material deformation and removal.

Table 3. The results of GRGA analysis on pile-up ratio.

Grey Comprehensive Relational Grade	Value	Rank
ρ_{01}	0.5680	3
ρ_{02}	0.5230	4
ρ_{03}	0.5962	2
ρ_{04}	0.7504	1

In summary, plastic deformation did not necessarily transform into pile-up when scratching Korean Pine. Most of the plastic deformation was absorbed by the anatomical cavities, especially in scratching Korean Pine longitudinally. And that is a marked difference from other engineered materials in sanding or grinding operations. In the case of scratching Korean Pine transversely, much cracked grain like fuzziness occurred owing to the shredding effect of side cutting-edges, and caused bad surface quality. Since early wood and late wood show diverse densities, there might be difference on material deformation and surface quality, especially in longitudinal direction. That difference was not taken into account for the scratches with gradual increased cutting depths generated in early wood or late wood randomly in this study. There is more pile-up on two sides of MDF scratch partly because MDF is a kind of densified and incompletely homogeneous material and the porosity is not significant. So, it can be inferred that the material porosity and wood grain direction are two key factors in the wood sanding processes.

3.2. The Elastic Spring Back Effect During Single Grit Grinding

Rubbing occurs at the initial stage of grit-workpiece interaction at a very small region, and only elastic deformation is included [12]. Since rubbing is not easily observed due to the elastic spring back effect and the three stages (rubbing, ploughing, cutting) along a scratch vary in proportion and can even occur simultaneously [34], only a few experimental researches on rubbing can be found in grinding metal, alloy or other engineered materials.

Wood is a kind of bio-based polymeric material, which can be regarded as elastic material when it is under small stress during a short period of time. Therefore, to investigate the surface creation mechanism thoroughly in scratching MDF and Korean Pine, the rubbing stage is necessary to be studied. In order to find out the elastic spring back effect during the scratching tests, a new method was developed by precisely matching the sanding force and the surface morphology. Specifically, the actual length (L') of the scratch can be measured by the analyzing software of the 3D profiler, and the grit entrance time (t_1) and exit time (t_2) of grit-workpiece interaction could be distinguished from the sanding force curve, as shown in Figure 10. Since the interaction time is very short with several microseconds and the penetration depth is relatively small, the velocity of the cutting-edge can be regarded as equal to the instant cutting speed v_c, which can be calculated as:

$$v_c = \frac{\pi D n}{60} \tag{7}$$

Then, the theoretical length (L) of the scratch can be calculated as:

$$L = v_c \times \Delta t \tag{8}$$

where $\Delta t = t_1 - t_2$.

During the arch trajectory, friction forces with the material must also be considered to see if the velocity decay is heavy. Here, the scratch of MDF in Figure 11 was taken as an example. The average tangential sanding force of this scratch is about 12 N, and the contact arch length was about 0.01 m, so the work of friction forces ΔE was approximately 0.12 J. If the friction energy loss was provided completely by the rotational kinetic energy of the wheel, then the velocity variation can be calculated as:

$$\frac{1}{2} J (\Delta v_c / R)^2 = \Delta E \tag{9}$$

where J is the moment of inertia of the wheel, and R is the wheel radius. So, the velocity decay is about 1.1 m/s. Moreover, the electrical spindle provided about 1.35 J during the Δt (0.9 ms). And that means the method proposed above is appropriate for this outstanding question.

The results are listed in Table 4. Compared with the actual length, the theoretical length was larger both in scratching MDF and Korean Pine ($\lambda = 0°$ and $\lambda = 90°$), which strongly indicates that the material elastic deformation recovered after the grit passes.

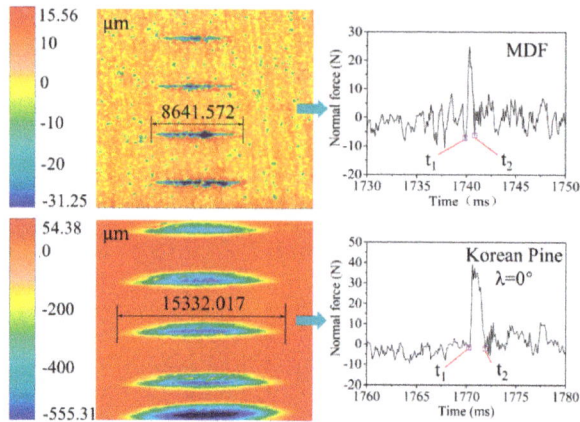

Figure 10. The accurate match of surface morphology and sanding force during a single scratch.

Table 4. Comparison of the theoretical length (*L*) and measured length (*L'*) of the scratch.

Comparison	*L* (μm)	*L'* (μm)	Δ*L* = *L* − *L'* (μm)
MDF	10489.5	8641.6	1847.9
Korean Pine (λ = 0°)	18648.0	15332.0	3316.0
Korean Pine (λ = 90°)	16317.0	12242.1	4074.9

In order to illustrate the accurate and comprehensive information on scratching Korean Pine (λ = 90°), the normal force voltage signal curve was plotted in Figure 11a. The elastic spring back effect seems to be more obvious at the exit side and a little longer path at the end of the scratch was observed, which is different from the case of λ = 0°. That is mainly because of the scratch direction. While the grit was leaving the workpiece from the deepest point, the fractured material accumulation in front of the rake face kept compressing the material around the grit cutting-edges, which caused elastic and plastic deformation simultaneously. In addition, the grit-workpiece contact length was decreasing due to the decreasing cutting depth, and the transverse fibre structure was hard to be severed when the cutting depth was relatively small. Therefore, the compressive action on the surface layer mainly led to elastic deformation and the elastic deformation recovered after the grit left the workpiece, as demonstrated in Figure 11b, which can perfectly account for the more obvious and stronger elastic spring back effect at the exit side of the scratch.

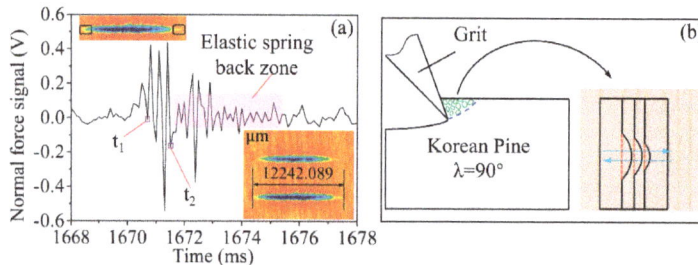

Figure 11. The analysis of elastic spring back effect when scratching Korean Pine (λ = 90°): (**a**) the exact match of sanding force and scratch morphology; (**b**) schematic of the elastic spring back effect when the grit was about to leave the workpiece.

4. Conclusions

The material removal and surface creation process of sanding MDF and Korean Pine was studied with single grit scratching tests. It is found that there are some differences in the material deformation during scratching Korean Pine and MDF. A mechanism was suggested to account for the differences. This mechanism helps to explain why tiny even no pile-up occurs during scratching Korean Pine, especially in the longitudinal direction. It can be inferred that the material porosity and wood grain direction are two key factors in the wood sanding processes. As for MDF, there is a decreasing tendency of pile-up ratio when the depth of cut and groove area increase, and the normal sanding force exerts the greatest influence on pile-up ratio. At the rubbing stage, a new method to confirm the elastic spring back effect was developed. And it is verified that elastic deformation both in MDF and Korean Pine scratching did exist through mathematical calculations. The material removal and surface creation are fertile areas for wood sanding research, and the approaches used here could provide insights to investigate other wood species and wood composites to improve sanding efficiency and surface quality eventually. Further research will focus on the effects of grit geometry and wood species with different densities and the effects of early wood and late wood on material removal and surface creation during the sanding process.

Author Contributions: Data Curation, J.Z.; Formal Analysis, J.Z.; Funding Acquisition, H.L.; Investigation, J.Y. and F.C.; Methodology, J.Z. and B.L.; Project Administration, L.L.; Software, J.Y. and F.C.; Supervision, B.L., H.L. and L.L.; Writing–Original Draft Preparation, J.Z.; Writing–Review & Editing, B.L., H.L. and L.L.

Funding: This research was funded by the Fundamental Funds for the Central Universities (No. 2017JC11).

Acknowledgments: The authors are grateful for the support of MOE Key Laboratory of Wooden Material Science and Application, Beijing Key Laboratory of Wood Science and Engineering at Beijing Forestry University.

Conflicts of Interest: The authors confirm that no conflict of interest exists in this article.

References

1. Ratnasingam, J.; Reid, H.; Perkins, M. The abrasive grinding of rubberwood (Hevea brasiliensis): An industrial perspective. *Eur. J. Wood Prod.* **2002**, *60*, 191–196. [CrossRef]
2. Bao, X.; Ying, J.H.; Cheng, F.; Zhang, J.; Luo, B.; Li, L.; Liu, H.G. Research on neural network model of surface roughness in belt grinding process for Pinus koraiensis. *Measurement* **2018**, *115*, 11–18. [CrossRef]
3. Sulaiman, O.; Hashim, R.; Subari, K.; Liang, C.K. Effect of grinding on surface roughness of rubberwood. *J. Mater. Process. Technol.* **2009**, *209*, 3949–3955. [CrossRef]
4. Stewart, H.A. Some surfacing defects and problems related to wood moisture content. *Wood Fiber* **1980**, *12*, 175–182.
5. Evans, P.H.; Cullis, I.; Kim, J.; Leung, H.L.; Hazneza, S.; Heady, R.D. Microstructure and mechanism of grain raising in Wood. *Coatings* **2017**, *7*, 135. [CrossRef]
6. Koehler, A. Some observations on raised grain. *Trans. Am. Soc. Mech. Eng.* **1932**, *54*, 27–30.
7. Taylor, J.B.; Carrano, A.L.; Lemaster, R.L. Quantification of process parameters in a wood sanding operation. *For. Prod. J.* **1999**, *49*, 41–46.
8. Luo, B.; Li, L.; Liu, H.; Xu, M.; Xing, F. Analysis of sanding parameters, sanding force, normal force, power consumption, and surface roughness in sanding wood-based panels. *Bioresources* **2014**, *9*, 7494–7503. [CrossRef]
9. Xu, M.; Li, L.; Wang, M.; Luo, B. Effects of surface roughness and wood grain on the friction coefficient of wooden materials for wood–wood frictional pair. *Tribol. Trans.* **2014**, *57*, 871–878. [CrossRef]
10. Cool, J.; Hernández, R.E. Improving the sanding process of black spruce wood for surface quality and water-based coating adhesion. *For. Prod. J.* **2011**, *61*, 372–380. [CrossRef]
11. Sun, Y.; Li, L. Effect on bonding strength of surface roughness of sanding wood. *Wood Process. Mach.* **2010**, *21*, 41–43. (In Chinese).
12. Hahn, R.S. On the Mechanics of the grinding process under plunge cut conditions. *J. Eng. Ind.* **1966**, *88*, 72–80. [CrossRef]

13. Klocke, F. Principles of cutting edge engagement. In *Manufacturing Process 2: Grinding, Honing, Lapping*; Kuchle, A., Ed.; Springer: Berlin/Heidelberg, Germany, 2009; pp. 3–16.

14. Chen, X.; Öpöz, T.T. Simulation of grinding surface creation—A single grit approach. *Adv. Mater. Res.* **2010**, *126–128*, 23–28. [CrossRef]

15. Anderson, D.; Warkentin, A.; Bauer, R. Comparison of spherical and truncated cone geometries for single abrasive-grain cutting. *J. Mater. Process. Technol.* **2012**, *212*, 1946–1953. [CrossRef]

16. Rasim, M.; Mattfeld, P.; Klocke, F. Analysis of the grain shape influence on the chip formation in grinding. *J. Mater. Process. Technol.* **2015**, *226*, 60–68. [CrossRef]

17. Öpöz, T.T.; Chen, X. Experimental investigation of material removal mechanism in single grit grinding. *Int. J. Mach. Tools Manuf.* **2012**, *63*, 32–40. [CrossRef]

18. Ohtani, T.; Tanaka, C.; Usuki, H. Comparison of the heterogeneity of asperities in wood and aluminum grinding surfaces. *Precis. Eng.* **2004**, *28*, 58–64. [CrossRef]

19. Gibson, L.J.; Ashby, M.F. *Cellular Solids: Structure and Properties*; Cambridge University Press: Cambridge, UK, 1988; p. 387.

20. Lee, P.H.; Nam, J.S.; Li, C.; Sang, W.L. An experimental study on micro-grinding process with nanofluid minimum quantity lubrication (MQL). *Int. J. Precis. Eng. Manuf.* **2012**, *13*, 331–338. [CrossRef]

21. Wang, S.; Li, C.; Zhang, X.; Zhou, D.; Zhang, D.; Zhang, Q. Modeling and simulation of the single grain grinding process of the nano-particle jet flow of minimal quantity lubrication. *Open Mater. Sci. J.* **2014**, *8*, 55–62. [CrossRef]

22. Widsten, P. Oxidative Activation of Wood Fibers for The Manufacture of Medium-Density Fiberboard (MDF). Ph.D. Thesis, Helsinki University of Technology, Espoo, Finland, November 2002.

23. Almeida, G.; Hernádez, R. Influence of the pore structure of wood on moisture desorption at high relative humidities. *Wood Mater. Sci. Eng.* **2007**, *2*, 33–44. [CrossRef]

24. Andersson, S. A study of the Nanostructure of the Cell Wall of the Tracheids of Conifer Xylem by X-ray Scattering. Ph.D. Thesis, University of Helsinki, Helsinki, Finland, January 2007.

25. Deng, J.L. Introduction to Grey system theory. *J. Grey Syst.* **1989**, *1*, 1–24.

26. Ding, W.D.; Koubaa, A.; Chaala, A.; Belem, T.; Krause, C. Relationship between wood porosity, wood density and methyl methacrylate impregnation rate. *Wood Mater. Sci. Eng.* **2008**, *3*, 62–70. [CrossRef]

27. Usta, I. Comparative study of wood density by specific amount of void volume (porosity). *Turk. J. Agric. For.* **2003**, *27*, 1–6.

28. De Moura, L.F.; Hernández, R.E. Effects of abrasive mineral, grit size and feed speed on the quality of sanded surfaces of sugar maple wood. *Wood Sci. Technol.* **2006**, *40*, 517–530. [CrossRef]

29. Tian, M.; Li, L. Study on influencing factors of sanding efficiency of abrasive belts in wood materials sanding. *Wood Res.* **2014**, *59*, 835–842.

30. Stewart, H.A.; Crist, J.B. SEM examination of subsurface damage of wood after abrasive and knife planing. *Wood Sci.* **1982**, *14*, 106–109.

31. De Meijer, M.; Thurich, K.; Militz, H. Comparative study on penetration characteristics of modern wood coatings. *Wood Sci. Technol.* **1998**, *32*, 347–365. [CrossRef]

32. Hernández, R.E.; de Moura, L.F. Effects of knife jointing and wear on the planed surface quality of northern red oak wood. *Wood Fiber Sci.* **2002**, *34*, 540–552.

33. De Moura, L.F.; Hernández, R.E. Evaluation of varnish coating performance for two surfacing methods on sugar maple wood. *Wood Fiber Sci.* **2005**, *37*, 355–366.

34. Öpöz, T.T.; Chen, X. Experimental study on single grit grinding of Inconel 718. *Proc. Inst. Mech. Eng. Part B J. Eng. Manuf.* **2014**, *229*, 713–726. [CrossRef]

coatings

MDPI

Article

The Effect of a Gear Oil on Abrasion, Scuffing, and Pitting of the DLC-Coated 18CrNiMo7-6 Steel

Remigiusz Michalczewski [1,*], Marek Kalbarczyk [1], Anita Mańkowska-Snopczyńska [1],
Edyta Osuch-Słomka [1], Witold Piekoszewski [1], Andrzej Snarski-Adamski [1], Marian Szczerek [1],
Waldemar Tuszyński [1], Jan Wulczyński [1] and Andrzej Wieczorek [2]

[1] Tribology Department, Institute for Sustainable Technologies–National Research Institute (ITeE-PIB),
   ul. K. Pulaskiego 6/10, 26-600 Radom, Poland; marek.kalbarczyk@itee.radom.pl (M.K.);
   anita.mankowska@itee.radom.pl (A.M.-S.); edyta.slomka@itee.radom.pl (E.O.-S.);
   witold.piekoszewski@itee.radom.pl (W.P.); andrzej.snarski@itee.radom.pl (A.S.-A.);
   marian.szczerek@itee.radom.pl (M.S.); waldemar.tuszynski@itee.radom.pl (W.T.);
   jan.wulczynski@itee.radom.pl (J.W.)
[2] Department of Mechanization and Robotics of Mining, Faculty of Mining and Geology, Silesian University
   of Technology, ul. Akademicka 2, 44-100 Gliwice, Poland; andrzej.n.wieczorek@polsl.pl
* Correspondence: remigiusz.michalczewski@itee.radom.pl; Tel.: +48-48-364-9247

Received: 14 November 2018; Accepted: 11 December 2018; Published: 20 December 2018

Abstract: The transmissions of mining conveyors are exposed to very harsh conditions. These are primarily related to the contamination of the gear oil with hard particles coming from coal and lignite, which can cause intensive abrasive wear, scuffing, and even pitting, limiting the life of gears. One of the ways to prevent this problem is the deposition of a wear-resistant coating onto gear teeth. However, a proper choice of gear oil is an important issue. The abrasion, scuffing, and pitting tests were performed using simple, model specimens. A pin and vee block tester was employed for research on abrasion and scuffing. To test pitting, a modified four-ball pitting tester was used, where the top ball was replaced with a cone. The test pins, vee blocks, and cones were made of 18CrNiMo7-6 case-hardened steel. A new W-DLC/CrN coating was tested. It was deposited on the vee blocks and cones. For lubrication, three commercial industrial gear oils were used: A mineral oil, and two synthetic ones with polyalphaolefin (PAO) or polyalkylene glycol (PAG) bases. The results show that, to minimize the tendency forabrasion, scuffing, and pitting, the (W-DLC/CrN)-8CrNiMo7-6 tribosystems should be lubricated by the PAO gear oil.

Keywords: DLC coating; gear oil; abrasive wear; scuffing; pitting

1. Introduction

Gears are very often exposed to very harsh conditions. This concerns gears used in the transmissions of chain and belt conveyors working in coal and open pit mines. The harsh conditions are primarily related to the contamination of the gear oil with hard particles coming from coal and lignite. This can cause intensive abrasive wear, scuffing, and even pitting, limiting the life of gears.

One of the ways to increase the life of gears in mining conveyors is the deposition of a thin, wear-resistant, low-friction coating onto gear teeth [1]. In the literature, one can find information on successful applications of various coatings to reduce the tendency of gears to scuffing [2–5], micropitting [6], and to possibly reduce friction [7,8]. However, another form of dangerous wear—pitting—may be accelerated when the coating is used [9–11]; although, in some cases, the resistance to pitting may be improved [12,13].

A variety of thin coatings have been tested. They can be divided into two groups: Non-DLC (DLC—diamond-like carbon) and DLC coatings. Tested non-DLC coatings are, for example, Nb–S [14],

MoS$_2$/Ti, C/Cr [4], TiN, and CrN [15]. Concerning DLC coatings, they are either doped coatings: W-DLC [2,15–17], Cr-DLC [18], and Si-DLC [19]; or non-doped coatings: a-C:H [18–20] or ta-C [19]. The review of the literature allows one to state that it is the DLC coating that is currently used in the majority of tribological research works.

Apart from testing thin coatings, it is also important to select a proper oil to lubricate the coated parts [15]. For decades, a lot of research works have been devoted to investigating the interaction between the lubricating additives in the oil with the steel surface [21–32], and the mechanisms of the interaction between steel surfaces and lubricants are well recognized.

Concerning the interaction between the oil and the thin coating, the publications are less frequent [15,18,33–38] and have been mostly issued in the last 20 years. Unlike the oil-steel interactions, when testing coatings, one can find different statements and observations in the literature. Some authors point out an effect of the coating's elemental composition, occurrence of the transfer of material between the samples, forming of protective films on the surface, or even chemical reactions of the coating with the lubricating additives in the oil.

In a review paper, Kalin et al. [33] compared oil-coating interactions when lubricating with oils with a mineral, synthetic ester, and polyalphaolefin (PAO) base using various tribosystems. They stated that non-doped DLC coatings can react with different types of additives (e.g., a friction modifier (FM), antiwear (AW), and extreme-pressure (EP) additives), and that the hydrogen content in DLC coatings plays a crucial role in the tribological performance under lubricated conditions.

Michalczewski et al. [15] performed four-ball scuffing tests, where all the balls were coated with a W-DLC coating, lubricated with a mineral oil with EP additives. After rubbing the coating off the upper ball, the researchers noticed the transfer of steel onto the coated surface of the lower balls, modified by the products of chemical reactions of sulfur and phosphorus coming from the EP additives.

Vlad et al. [34] performed four-ball scuffing tests, where the upper ball was coated with a W-DLC coating, lubricated by a polyalphaolefin (PAO) oil with an AW additive. They observed the transfer of the coating onto the steel counter-face of the bottom balls and the reaction of the AW additive with the steel surface, generating a protective film, consisting of phosphates, oxides, and sulfides as reaction products.

Mistry et al. [35] performed ball-on-disc tests with a W-DLC coating deposited on both test specimens, lubricated with PAO oil with AW and EP additives. On the worn surface, they identified wear-reducing compounds (zinc sulfide and tungsten carbide) as well as friction reducing compounds (e.g., tungsten sulfide), which formed as a result of coating-lubricant interactions.

Vengudusamy et al. [19] conducted a test on the non-doped and doped DLC coatings deposited on the test specimens of a minitraction machine (MTM), lubricated with a mineral base oil. They stated that one of the reasons for the reduction in friction may be the formation of a thin oxygen-containing layer. On the other hand, according to Yang et al. [37], performing pin-on-flat experiments of the W-DLC/cast iron tribosystem, lubricated with a PAO oil with AW and FM additives, the lubricant additives reduce the formation of tungsten oxides, which is a more brittle material, that could lead to the failure of the coating, as is seen in base oil lubrication.

Yang et al. [38] conducted reciprocating pin-on-plate experiments on a W-DLC/cast iron tribosystem lubricated with an oil withAW and FM additives. They stated that the formation of MoS$_2$, by chemical decomposition from MoDTC, is dominant in such a system, rather than the possible formation of WS$_2$. Similar observations were done by Haque et al. [20] by performing reciprocating pin-on-plate experiments of the non-doped DLC/cast iron tribosystem, lubricated with a PAO oil with AW and FM additives. They observed the formation of a low friction MoS$_2$ sheet and ZnO/ZnS compounds at the interface, as well as the formation of a ZDDP tribofilm containing Zn/ZnO/ZnS.

The most often used lubrication in tribosystems with DLC coatings are oils with a mineral or PAO base. Concerning increasingly used polyalkylene glycol (PAG) oils, there are only a few papers on their interaction with a DLC coating (e.g., [39]).

In most of the works referred to above, lubricating oils were "model oils" (i.e., they were formulated by the researchers to control their chemical composition and more precisely identify the surface phenomena). However, where an application of the coatings to increase the life of gears is at stake, it is necessary to use commercial, fully-formulated gear oils, having the performance level and viscosity grade close to the oils used in specific applications. Only a minority of the research works concern the use of fully-formulated, commercial oils (e.g., [1,15,40–42]).

In the research and development works, before verification (component) tests are performed (e.g., on gears), screening tests on model, simple specimens are very often carried out. This approach is suggested (e.g., in the works [43–47]). The reason is that component (gear) tests are very expensive and time-consuming. For example, gear pitting tests may require time counted in months. This is why the majority of tribological experiments on thin coatings are carried out using model, simple specimens, especially when testing pitting [15,48–51].

This paper presents an effect of fully formulated gear oils, intended for the lubrication of gears made of 18CrNiMo7-6 steel (where one gear will be DLC-coated) on abrasion, scuffing, and pitting. The gear oils used had three types of base oils: Mineral, PAO, and PAG. The tests were performed using model, simple specimens. A new, commercially available W-DLC/CrN coating was tested, developed from the point of view of reducing the negative impact of high cyclic stress.

The final aim of the research is to improve the durability of the planetary transmissions in mining conveyors.

2. Materials and Methods

For research on abrasive wear and scuffing, the T-09 pin and vee block tribotester was employed. This device was developed and manufactured at the Institute for Sustainable Technologies–National Research Institute in Radom, Poland.

The tested tribosystem is shown in Figure 1.

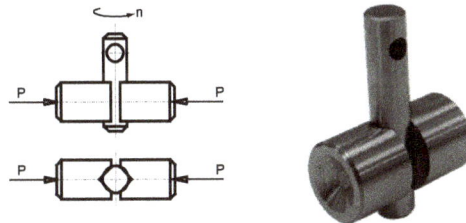

Figure 1. Pin and vee block tribosystem.

The tribosystem consisted of two vee blocks pressed at a certain load P against the test pin that rotated at the constant speed of n (290 rpm), driven by the test shaft through the shear pin inserted in the hole. The shear resistance of all the shear pins was almost identical. The friction contact was lubricated by the oil poured into the reservoir equipped with a heater, where the contact area was immersed. The initial temperature of the oil was set up at the level of 70 °C, which corresponds to the temperature expected in the transmissions of mining conveyors.

The abrasion test was carried out according to ASTM D2625 [52], Procedure A, slightly modified. First, the run-in procedure was performed—testing at 1334 N for 3 min. Then, the regular test was carried out. In the first phase, the load of 2224 N was applied for 1 min or up to failure. If no failure was observed, the load was increased to 3336 N, and the test was performed for the next minute or up to failure. If no failure was observed, the load was again increased to 4448 N and the test was performed until failure.

Failure was indicated by a sharp torque rise of 1.13 Nm above the steady-state value, or the breakage of the shear pin, or the inability to maintain the load, or exceeding 10,000 s of the total time excluding the 3-min run-in (i.e., 164 min).

The resistance to abrasion was expressed by the endurance (wear) life, which is the total time of the runs before failure occurs, excluding the 3-min run-in. It was adopted that a minimum of 3 test runs constitute the final result.

The scuffing test was carried out according to ASTM D3233 [53], Method A. First, the run-in procedure was performed—testing at 1334 N for 5 min. Then, the regular test was carried out—the load was continuously increased up to failure or until the maximum load was attained.

Failure was indicated by breakage of the shear pin or the test pin.

The resistance to scuffing was expressed by the load at failure or a maximum attainable load. It was adopted that a minimum of 4 test runs constitute the final result.

A comparison of the load change in both tests is shown in Figure A1. A continuous load increase (Figure A1b) gives very harsh test conditions and differentiates scuffing tests from abrasion tests.

For research on pitting, the T-02U universal four-ball testing machine was employed. This device was developed and manufactured at the Institute for Sustainable Technologies–National Research Institute in Radom, Poland. In the tribosystem the top ball was replaced with a cone [51].

The tested tribosystem and the SEM image of a typical pit on the cone are shown in Figure 2.

(a)   (b)

Figure 2. (**a**) Cone-balls tribosystem:1—cone, 2—balls, 3—race; (**b**) SEM image of a typical pit on the cone.

Figure 2a shows that the tribosystem consisted of three balls (2) that were free to rotate in the special race (3) and were pressed at the required load P against the cone (1). The cone was fixed in the ball chuck and rotated at the defined speed n. The contact zone of the balls was immersed in the oil. The holder of the race (3) was equipped with a heater. As in the case of the previously described tests, the initial temperature of the oil was set up at the level of 70 °C.

The pitting test was performed according to IP 300 standard [54]. Test conditions were as follows: Rotational speed of 1450 rpm, applied load of 3924 N (400 kgf), run duration until pitting occured, and the number of runs of 24. Only those runs for which pitting occurred on the cone were accepted. In each run, the time to pitting failure occurrence was measured.

After test completion, the 24 values (failure times) were plotted in Weibull co-ordinates (i.e., the estimated cumulative percentage failed against the failure time). Then, a straight line was fitted to the points. From the line, the fatigue life L_{10} was read off, which expresses the resistance to pitting. The value of L_{10} represents the life at which 10% of a large number of test cones would be expected to have failed.

The test pins, vee blocks, and cones were made of 18CrNiMo7-6 case-hardened steel, which is the material intended for the gears in mining conveyors. The hardness was 62 HRC. The roughness of the vee blocks and cones was $R_a = 0.20$ μm, and the pins was $R_a = 0.52$ μm. The balls and races (pitting tests) were made of 100Cr6 bearing steel.

The W-DLC/CrN, commercially available, antifriction coating was tested. The coating was deposited by reactive sputtering in the physical vapor deposition (PVD) process. One kind of specimen was coated (e.g., on the vee blocks (abrasion and scuffing tests) and on the cones (pitting tests)), leaving the counter-specimens uncoated. The reason for this is that, in the planetary transmissions of mining conveyors, only planetary pinions will be coated, leaving the ring and sun gears uncoated.

The microstructure of the coating revealed during qualitative glow-discharge optical emission spectrometry (GDOES)analysis is shown in Figure 3.

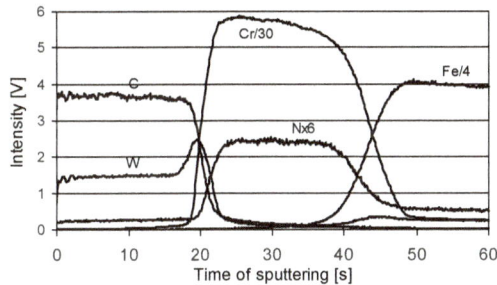

Figure 3. GDOES depth profile of the W-DLC/CrN coating.

The W-DLC/CrN coating represents an a-C:H:Me group. Figure 3 shows that the coating consists of CrN adhesion layer adjacent to the steel substrate, followed by the WC interlayer, and then the WC/C layer (i.e., alternating lamellae of WC and a hydrocarbon layer doped with tungsten). WC/C layer exhibits antifriction properties. The producer of such coatings states that the friction coefficient at dry friction, when rubbing against a steel ball, is between 0.1 and 0.2. When the WC/C layer is worn away, the hard WC interlayer is exposed. This is followed by the exposure of CrN, and these layers have very good resistance to abrasion. The CrN layer is very ductile, and it additionally gives excellent support properties in cyclic loaded contacts. Actually, in the preliminary experiment, the coating used in this research gave a much better resistance to pitting than another DLC coating that was deposited on the adhesive layer of only chromium.

The basic properties of the W-DLC/CrN coating were the following: A thickness of 4.2 m, an adhesion in scratch tests of 151 N, a nanohardness of 12 GPa, a roughness R_a of 0.09 μm, and a roughness R_z of 1.04 μm. The thicknesses of the WC/C, WC, and CrN layers were the following, respectively: 1.9, 0.4, and 1.9 μm.

For lubrication, commercial industrial gear oils were used. They were a mineral oil, and two synthetic ones (i.e., one with a polyalphaolephine (PAO) base and one with a polyalkylene glycol (PAG) base, and all with the viscosity grade of VG 320. The most important difference is in theviscosity index (VI). For the mineral oil, VI = 95, for the PAO oil VI = 159, and for the PAG oil VI = 230. Such oils are used, for example, to lubricate the gears of mining conveyors.

To determine the features of the worn surface, the following analytical instruments were used: A field-emission scanning electron microscope (FE-SEM, SU-70, Hitachi, Tokyo, Japan), an energy dispersive spectrometer (EDS, Noran System 7, Thermo Scientific, Waltham, MA, USA), an optical profilometer (Talysurf CCI, Taylor Hobson, Leicester, UK), a stylus profilometer (Talysurf PGI 830, Taylor Hobson), an atomic force microscope (AFM, Q-Scope 250, Quesant Instrument Corporation, Agoura Hills, CA, USA), an optical microscope (MM-40, Nicon, Tokyo, Japan), and a microhardness meter (FM-800, Future-Tech Corp., Kawasaki, Japan). Prior to the microanalyses, the test specimens were washed with 95% *n*-hexane in an ultrasonic washer for 10 min, and then dried in the open air.

In reference to the EDS analysis, it was either qualitative analysis (surface maps for particular elements) or quantitative analysis(point analysis). Both types of analyses were performed at the accelerating voltage of 15 kV. For the quantification, the widely used correction method "Phi-Rho-Z" was employed. The quantitative analysis had no standards.

Concerning statistical analysis, for the results of the abrasion and scuffing tests, confidence intervals representing 95% probability were calculated. For the results of the pitting tests, confidence intervals representing 90% probability were calculated.

3. Results

3.1. Abrasion Tests

The results obtained in the abrasion tests for the three gear oils are presented in Figure 4. The type of the bases of the gear oils are denoted as M (mineral), PAO, and PAG. For reference, the results for the uncoated specimens lubricated with the mineral oil are also shown.

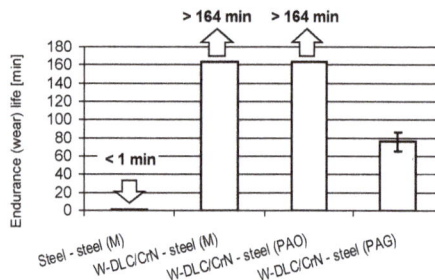

Figure 4. Endurance (wear) lives obtained in abrasion tests.

As can be observed from Figure 4, for the three gear oils lubricating the (W-DLC/CrN)-steel tribosystem, the resistance to abrasion notably rose in comparison with the reference specimens (i.e., the steel-steel tribosystem lubricated with the mineral oil). The mineral and PAO oils gave the same results. PAG oil gave the worst result.

As was observed, in all the tests on the uncoated tribosystem and the tribosystem lubricated with the PAG oil, the breakage of the shear pin took place due to a high friction torque. This suggests the occurrence of scuffing in the test called the "abrasion test", and in turn difficulties in separation of abrasion and scuffing, which may appear under the same test conditions depending on the tribosystem materials and lubricant. However, one needs to have in mind that the standard test, performed according to ASTM D2625 [52], Procedure A, is dedicated to testing the endurance (wear) life of dry solid film lubricants (i.e., dry coatings consisting of lubricating powders in a solid matrix bonded to one or both surfaces to be lubricated). Such coatings are low friction ones, similarly to the external WC/C layer of the coating used in this research. That means that, even in the abrasion tests, "undesired" scuffing, which may appear only in their final phase, indicates an increase in friction due to the wearing away of the entire low friction coating (e.g., solid film lubricants) or its external low friction layer (e.g., WC/C in this research). The term "abrasion" adopted in the text of this paper should be treated as "endurance (wear) life". Thus, the used methodology is correct.

Optical profilometric images of the wear scars on the vee blocks are shown in Figure 5.

In the case of the uncoated tribosystem, the depth of the wear scar is the biggest. For the coated vee blocks, the wear scars are much shallower. In the case of the lubrication with the mineral oil, the outer layers of the W-DLC/CrN coating were worn away, namely, the low-friction layer of WC/C and the hard, wear resistant interlayer of WC (having a total thickness of 2.3 µm), exposing the adhesive layer of CrN. As will be proved in the next section of the paper, for the PAO and PAG oils the outer layers were partially worn.

The average depths of the wear tracks on the test pins and surface roughness values of the worn area are compiled in Table 1.

The smallest depth of the wear track was observed for the uncoated tribosystem (because of the shortest run duration) and the tribosystem with the coated vee block, lubricated with the PAO oil. When comparing the initial roughness of the test pin with the roughness of the worn area, one can observe that the pin was slightly polished by the mild, abrasive action of the hard coating having numerous droplets on the surface. Figure 6 presents an AFM image of the surface of the W-DLC/CrN coating.

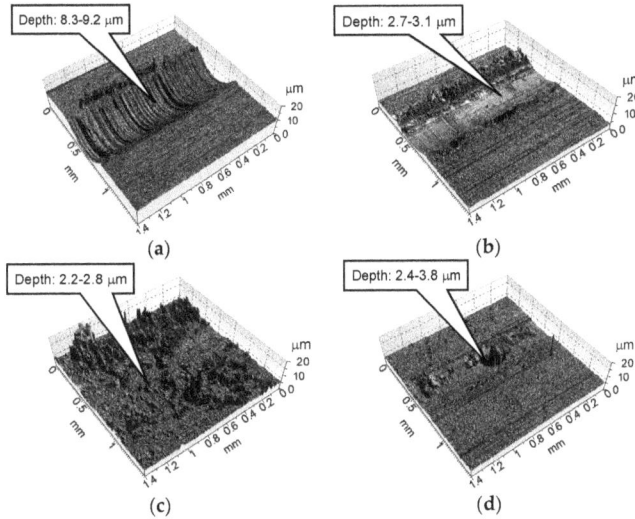

Figure 5. Optical profilometric images of the wear scars on the vee blocks: (**a**) Uncoated vee block lubricated with the mineral oil; (**b**) (W-DLC/CrN)-coated vee block lubricated with the mineral oil; (**c**) (W-DLC/CrN)-coated vee block lubricated with the PAO oil; and (**d**) (W-DLC/CrN)-coated vee block lubricated with the PAG oil.

Table 1. The average depths of the wear tracks on the test pins and surface roughness of the worn area measured by R_a and R_z parameters—stylus profilometry measurements.

Tested Tribosystem	Depth (µm)	R_a (µm)	R_z (µm)
Steel pin—steel vee block; mineral oil	≈0	0.41	2.99
Steel pin—(W-DLC/CrN)-coated vee block; mineral oil	0.8	0.21	1.42
Steel pin—(W-DLC/CrN)-coated vee block; PAO oil	≈0	0.15	1.01
Steel pin—(W-DLC/CrN)-coated vee block; PAG oil	0.5	0.25	1.74
Average initial roughness of the test pins	–	0.52	2.82

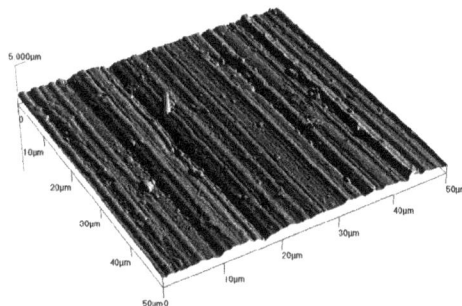

Figure 6. AFM image of the surface of the W-DLC/CrN coating.

From Table 1, it is apparent that, when the coated vee block was lubricated with the PAO oil, the worn area of the test pin was the smoothest.

3.2. Scuffing Tests

The results obtained in the scuffing tests for the three gear oils are presented in Figure 7. For reference, the results for the uncoated specimens lubricated with the mineral oil are also shown.

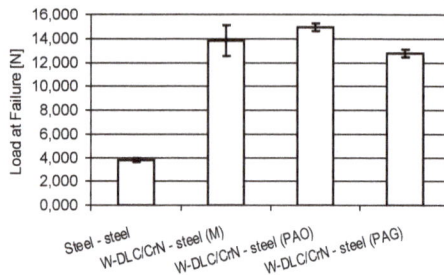

Figure 7. Loads at failure obtained in the scuffing tests.

As can be observed from Figure 7, for the three gear oils lubricating the (W-DLC/CrN)-steel tribosystem, the resistance to scuffing significantly rose in comparison with the reference specimens. The highest resistance to scuffing was given by the mineral and PAO oils. Similar to the behavior during the abrasion tests, PAG oil gave the worst result.

Optical profilometric images of the wear scars on the vee blocks are shown in Figure 8.

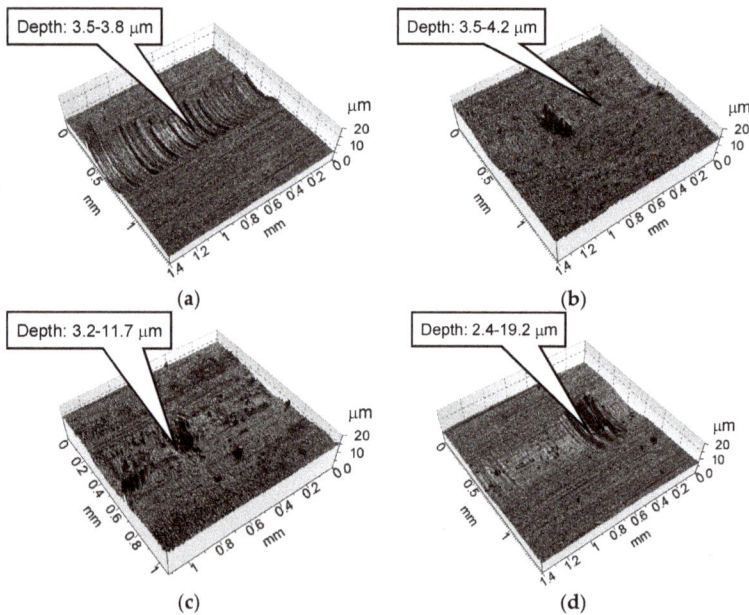

Figure 8. Optical profilometric images of the wear scars on the vee blocks: (**a**) Uncoated vee block lubricated with the mineral oil; (**b**) (W-DLC/CrN)-coated vee block lubricated with the mineral oil; (**c**) (W-DLC/CrN)-coated vee block lubricated with the PAO oil; and (**d**) (W-DLC/CrN)-coated vee block lubricated with the PAG oil.

The depths of the wear scars were similar. In the case of the lubrication with the mineral oil, the outer layers of the W-DLC/CrN coating were worn away, namely, the low-friction layer of WC/C and the hard, wear-resistant interlayer of WC (having a total thickness of 2.3 μm), exposing the adhesive layer of CrN. As will be proved in the next section of the paper, for the PAO and PAG oils, the layer of WC was not worn away. However, in these cases, local removal of the entire coating took place, giving very deep "craters" (Figure 8).

The average depths of the wear tracks on the test pins and the surface roughness of the worn area are compiled in Table 2.

Table 2. The average depths of the wear tracks on the test pins and the surface roughness of the worn area, measured by R_a and R_z parameters—stylus profilometry measurements.

Tested Tribosystem	Depth (μm)	R_a (μm)	R_z (μm)
Steel pin—steel vee block; mineral oil	≈0	0.28	1.73
Steel pin—(W-DLC/CrN)-coated vee block; mineral oil	≈0	0.16	1.15
Steel pin—(W-DLC/CrN)-coated vee block; PAO oil	1.2	0.13	0.84
Steel pin—(W-DLC/CrN)-coated vee block; PAG oil	≈0	0.32	2.80
Average initial roughness of the test pins	–	0.52	2.82

The smallest depth of the wear track was observed for the uncoated tribosystem (because of the shortest run duration) and the tribosystem with the coated vee block lubricated with the mineral and PAG oils. When comparing the initial roughness of the test pin with the roughness of the worn area, one can observe that the pin was slightly polished by the mild, abrasive action of the hard coating having numerous droplets on the surface (Figure 6). However, similar polishing was also observed for the uncoated tribosystem.

From Table 2, it is apparent that, when the coated vee block was lubricated with the PAO oil, the worn area of the test pin was the smoothest, similarly to what took place in the abrasion tests.

3.3. Pitting Tests

The results obtained in the pitting tests for the three gear oils are presented in Figure 9. For reference, the results for the uncoated specimens lubricated with the mineral oil are also shown.

As can be observed from Figure 9, when the coating was deposited on one specimen, the resistance to pitting dropped significantly, regardless of the oil used. It is also shown that the mineral and PAG oils exhibited similar resistance to pitting. The PAO oil gave a slightly worse result.

SEM images of the wear tracks on the test cones are shown in Figure 10.

In the case of the uncoated tribosystem, the depth of the wear track was the smallest. For the coated cones, the wear tracks were much deeper. As will be proved in the next section of the paper, irrespective of the oil used, the outer layers of the W-DLC/CrN coating were worn away, namely, the low-friction layer of WC/C and the hard, wear-resistant interlayer of WC (having a total thickness of 2.3 μm), exposing the adhesive layer of CrN. Very big depths of the wear tracks on the coated cones were a result of plastic deformation of the substrate material. However, the presence of a CrN layer is possible, because it is very ductile and adapts easily to the plastically deformed steel.

To sum up, taking into consideration various criteria (e.g., the resistance to abrasion, scuffing, wear, and roughness of the test specimens), the PAO gear oil was well suited for the lubrication of the (W-DLC/CrN)-18CrNiMo7-6 steel tribosystems. Although this oil gave a slightly worse resistance to pitting than the other oils lubricating the coating-steel contact area, the decisive reason for such a statement is that PAO exhibits a much better viscosity index than mineral oils, which is desired in the very extreme working conditions of mining conveyors.

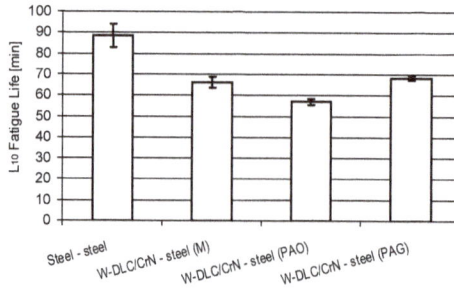

Figure 9. L_{10} fatigue lives obtained in the pitting tests.

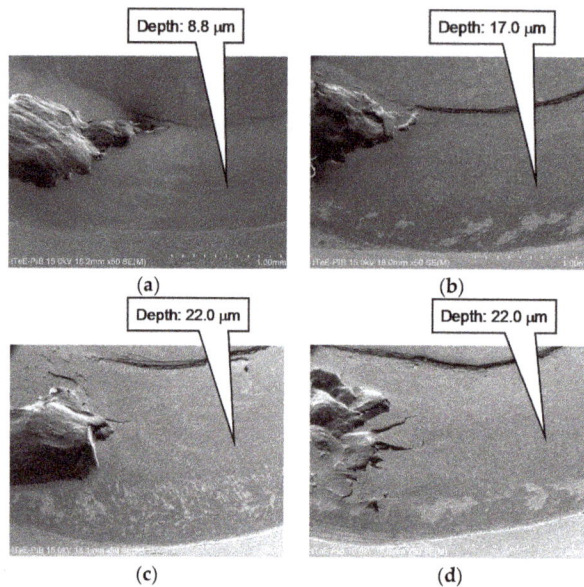

Figure 10. SEM images of the wear tracks on the test cones: (**a**) Uncoated cone lubricated with the mineral oil; (**b**) (W-DLC/CrN)-coated cone lubricated with the mineral oil; (**c**) (W-DLC/CrN)-coated cone lubricated with the PAO oil; and (**d**) (W-DLC/CrN)-coated cone lubricated with the PAG oil.

4. Discussion

4.1. Means of Analyses

To interpret the tribological results, surface elemental analyses and physico-mechanical analyses of the worn areas were conducted.

Figure A2 shows a photograph of the tested vee blocks and pin. Figure A3 presents a photograph of the tested cones. In Figures A2 and A3, white rectangles mark the places of surface elemental analyses.

4.2. Abrasion Tests

Figure 4, presented earlier, shows that the resistance to abrasion of the coating-steel tribosystem was much higher than for steel-steel, irrespective of the oil used. This is due to the action of the hard

WC interlayer in the coating, and when WC is worn away, the CrN layer gives very good resistance to abrasion. However, the question is, "why did the PAG oil produce the worst result?"

Results of SEM/EDS surface analysis from the worn surfaces of the vee blocks is shown in Figure 11.

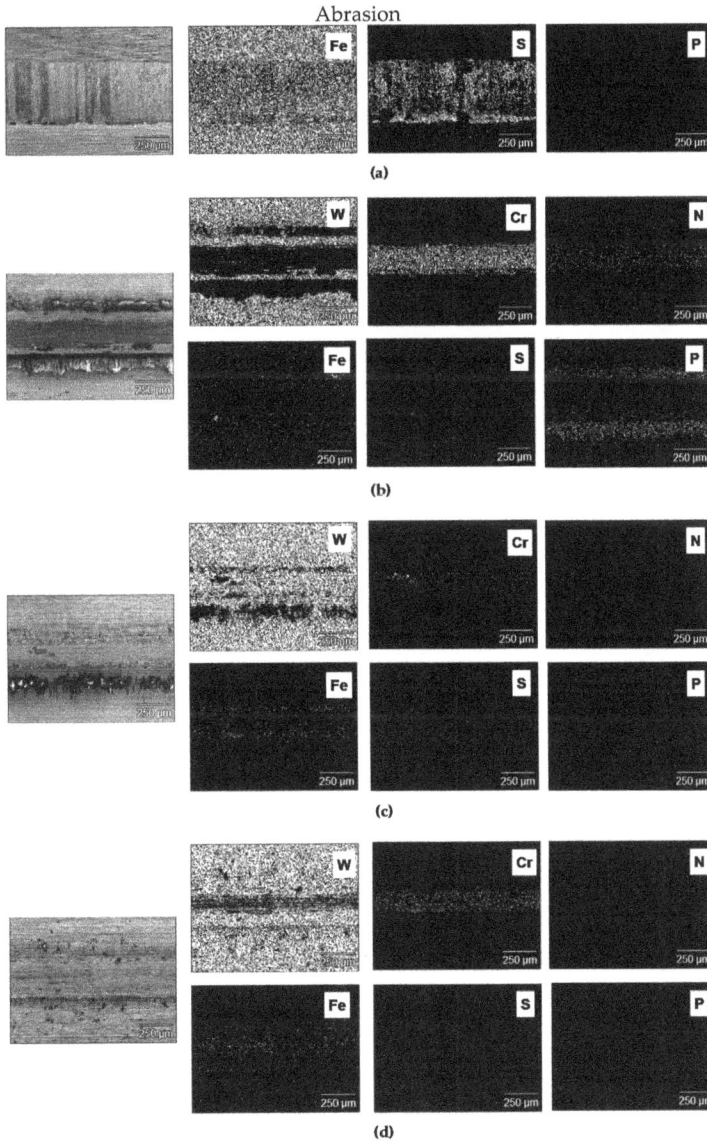

Figure 11. SEM images (on the left) and EDS surface analyses of the wear scars on the vee blocks: (**a**) Uncoated vee block lubricated with the mineral oil; (**b**) (W-DLC/CrN)-coated vee block lubricated with the mineral oil; (**c**) (W-DLC/CrN)-coated vee block lubricated with the PAO oil; and (**d**) (W-DLC/CrN)-coated vee block lubricated with the PAG oil.

In the case of the lubrication with the mineral oil, the outer layers of the W-DLC/CrN coating were worn away, namely, the low-friction layer of WC/C and the hard, wear-resistant interlayer of WC (having a total thickness of 2.3 μm), exposing the adhesive layer of CrN. For the PAO and PAG oils, the outer layers were partially worn. The coating removal was observed also in the earlier work by Michalczewski et al. [15].

As in the above cited work [15], the transfer of iron from the steel pin to the coated surface of the vee block due to friction was observed; although, in the case when the coated vee block was lubricated with the mineral oil, the transferred material was located mainly on the verges of the wear scar. The production of, for example, tungsten sulfide may be excluded, as suggested elsewhere (e.g., by Yang et al. [38]), as the places of appearance of sulfur did not "overlap" the areas where tungsten was identified, as in the case of the PAO oil. Sulphur was, rather, "in line" with the traces of iron, which suggests that iron sulfide (FeS)was produced on the transferred layer of iron, or that this compound was transferred from the chemically changed surface of the test pin due to the chemical reactions of the EP additives in the oil with the steel.

Vlad et al. [34] observed the transfer of the coating onto the steel counter-face. Thus, a possible transfer of the coating, identified by observing tungsten, to the surface of the test pin was also investigated. Table 3 compiles the results of the EDS analysis of tungsten on the worn surfaces of the pins.

Table 3. The traces of tungsten on the worn surfaces of the test pins.

Tested Tribosystem	Concentration of Tungsten (wt.%)
Steel pin—steel vee block; mineral oil	0.22–0.30
Steel pin—(W-DLC/CrN)-coated vee block; mineral oil	0.32–0.87
Steel pin—(W-DLC/CrN)-coated vee block; PAO oil	0.03–0.47
Steel pin—(W-DLC/CrN)-coated vee block; PAG oil	0.60–0.82

The results of the concentration of tungsten on the worn surface of a test pin working in the uncoated tribosystem may be considered to be a "background" signal. For the all tribosystems with the coated vee block, the maximum concentration of tungsten in the surface layer of the pins was higher than the background signal. This suggests the transfer of the external, low-friction layer of the coating, namely WC/C, to the pin surface, lowering the friction in this way and improving the results in the abrasion tests when the coated vee blocks were used. However, it does not explain the worse results obtained for the PAG oil.

Results of the quantitative EDS analysis of the sulfur and phosphorus concentration in the worn surfaces of the vee blocks and pins are shown in Figures 12 and 13, respectively. Each analysis was performed three times at different points. The scatters between the minimum and maximum values were added to the graphs.

The scatters of results showed a very inhomogeneous concentration of sulfur and phosphorus in the wear surface. The exception was the coating-steel tribosystem lubricated with the PAG oil.

It is known that the EP (extreme-pressure) additives (based on organic S–P compounds) form, for example, iron sulfide (FeS) [21]. FeS compounds, apart from hampering the creation of adhesive bonds, with their five-fold lower shear strength and four-fold lower hardness than steel, facilitate shearing of the chemically modified surface asperities. The shear plane is transferred to the thin FeS layer, which protects the surface from tearing out the material from deeper layers, reducing the wear intensity. Under extreme pressure conditions, a dominating role is played by products of the chemical reaction of sulfur [55].

In the steel-steel tribosystem, both on the worn surface of the vee blocks and of the pins, there were products of the reaction of sulfur, especially, with the steel surface—much more than when the coating was tested. In spite of this, the resistance to the abrasion of the coating-steel tribosystem was much higher than for steel-steel, irrespective of the oil used. As stated before, this is due to the action of the hard WC interlayer in the coating, and when WC is worn away, the CrN layer gives very good resistance to abrasion. Thus, the presence of the coating in the contact zone is of much greater significance than the chemical reactions with the oil.

Sulfur compounds found on the worn surface of the coated vee blocks were the result of chemical reactions of EP additives in the oil with the steel material, which was transferred from the test pin onto the vee block, and the transfer of chemically modified steel from the pin. Such a transfer was also observed in the earlier work [15]. As stated by other researchers (e.g., Mistry et al. [35] or Haque et al. [20]), the appearance of sulfur compounds on the worn surfaces of the test specimens adds, beneficially, to the improvement of the tribological features. For the PAG oil, the concentration of sulfur in the worn area of the both test specimens was the lowest. It was accompanied by the relatively low concentration of phosphorus. This is a reason for the lowest resistance to abrasion, in this case, among the coating-steel tribosystems.

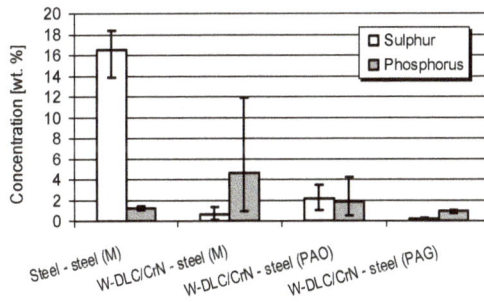

Figure 12. Quantitative results of the EDS analysis of the sulfur and phosphorus from the worn surfaces of the vee blocks.

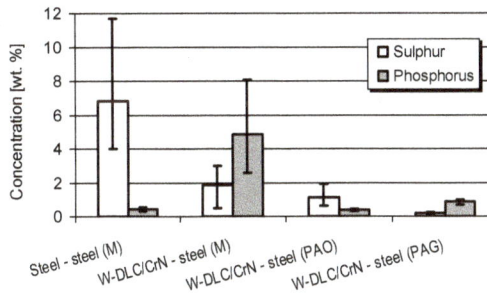

Figure 13. Quantitative results of the EDS analysis of the sulfur and phosphorus from the worn surfaces of the pins.

4.3. Scuffing Tests

Figure 7, presented earlier, shows that the resistance to scuffing of the coating-steel tribosystem was much higher than for steel-steel, irrespective of the oil used. The improvement of the resistance to scuffing was also observed by Kalin and Vižintin [2], Martins et al. [3,4], and by the authors of this paper [5] in case of the coated gears.

In this research, the improvement of the scuffing resistance was due to the action of the hard WC interlayer in the coating, and when WC was worn away, the CrN layer gave very good resistance to abrasion. A lower affinity of two different rubbing materials, than in the case of the uncoated contact, as well as the high hardness of the WC and CrN layers in the coating, led to a smaller tendency for the creation of adhesive bonds, which gave a lower tendency for scuffing. However, as in the abrasion tests, the question is, "why did the PAG oil produce the worst result?"

Results of SEM/EDS surface analysis from the worn surfaces of the vee blocks is shown in Figure 14. Because, in the scuffing tests, in the case of lubrication by the PAO and PAG oils, the coating was unevenly worn along the wear scar, the analyses came from the dominating regions, in relation to the wear depth.

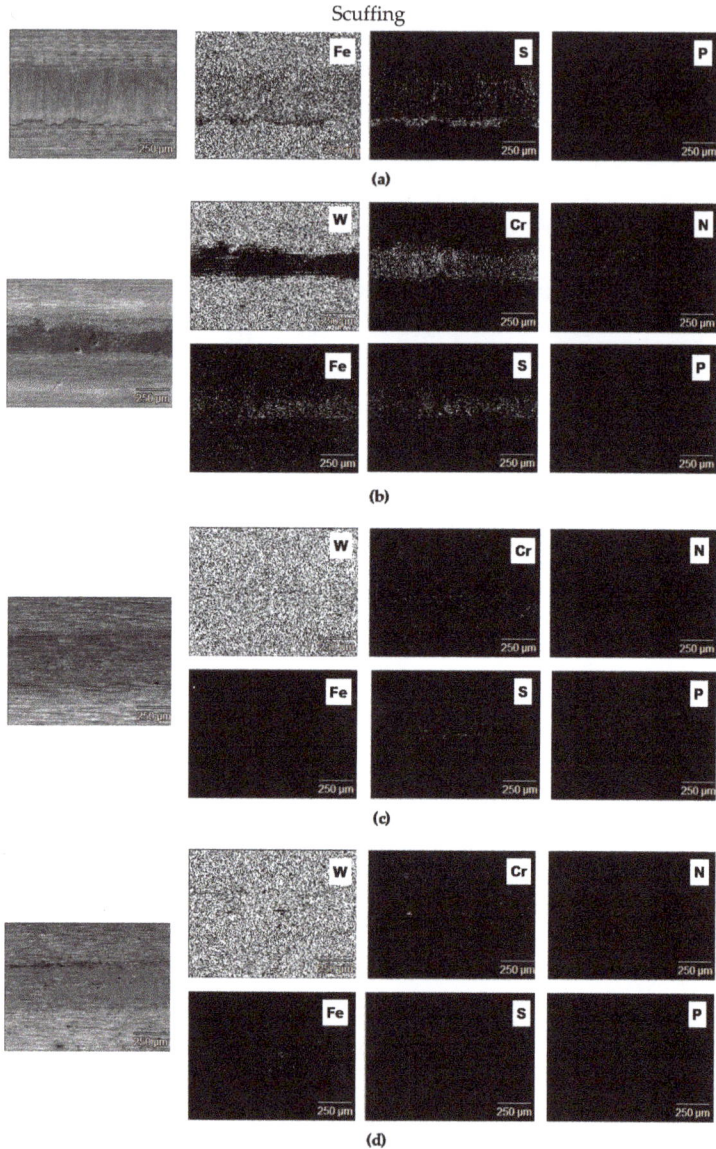

Figure 14. SEM images (on the left) and EDS surface analyses of the wear scars on the vee blocks: (**a**) Uncoated vee block lubricated with the mineral oil; (**b**) (W-DLC/CrN)-coated vee block lubricated with the mineral oil; (**c**) (W-DLC/CrN)-coated vee block lubricated with the PAO oil; and (**d**) (W-DLC/CrN)-coated vee block lubricated with the PAG oil.

In the case of the lubrication with the mineral oil, the outer layers of the W-DLC/CrN coating were worn away, namely, the low-friction layer of WC/C and the hard, wear-resistant interlayer of WC (having a total thickness of 2.3 µm), exposing the adhesive layer of CrN. For the PAO and PAG oils, the outer layers were partially worn. The coating removal was also observed in the earlier work by Michalczewski et al. [15].

The transfer of iron from the steel pin to the coated surface of the vee block due to friction was observed only in the case when the coated vee block was lubricated with the mineral oil. The production of, for example, tungsten sulfide, may rather be excluded, as suggested elsewhere (e.g., by Yang et al. [38]), as the places of the appearance of sulfur did not "overlap" the areas where tungsten was identified, as in the case of the mineral oil lubricating the tribosystem with the coated vee block. Sulfur was, rather, "in line" with the traces of iron, which suggests that iron sulfide (FeS) was produced on the transferred layer of iron, or that this compound was transferred from the chemically changed surface of the test pin due to the chemical reactions of the EP additives in the oil with the steel.

As mentioned earlier, Vlad et al. [34] observed the transfer of the coating onto the steel counter-face. Thus, a possible transfer of the coating, identified by observing tungsten, to the surface of the test pin was also investigated. Table 4 compiles the results of the EDS analysis of tungsten on the worn surfaces of the pins.

Table 4. The traces of tungsten on the worn surfaces of the test pins.

Tested Tribosystem	Concentration of Tungsten (wt.%)
Steel pin—steel vee block; mineral oil	0.36–0.79
Steel pin—(W-DLC/CrN)-coated vee block; mineral oil	0.34–0.66
Steel pin—(W-DLC/CrN)-coated vee block; PAO oil	0.79–1.54
Steel pin—(W-DLC/CrN)-coated vee block; PAG oil	1.94–9.99

The results of the concentration of tungsten on the worn surface of the test pin working in the uncoated tribosystem may be considered to be a "background" signal. For the tribosystems with the coated vee block lubricated with the PAO and PAG oils, the maximum concentration of tungsten in the surface layer of the pins was bigger than the background signal. This suggests the transfer of the external, low-friction layer of the coating, namely WC/C, to the pin surface, lowering the friction in this way and improving the results in the scuffing tests when the coated vee blocks were used. However, as in the abrasion tests, it does not explain the worst result obtained for the PAG oil.

Results of the quantitative EDS analysis of the sulfur and phosphorus concentration in the worn surfaces of the vee blocks and pins are shown in Figures 15 and 16, respectively. Each analysis was performed three times at different points. The scatters between the minimum and maximum values were added to the graphs.

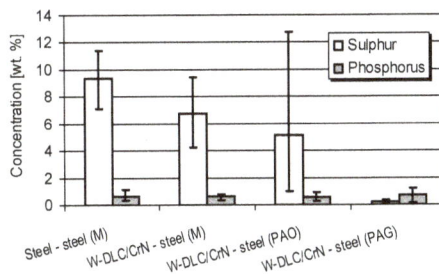

Figure 15. Quantitative results of the EDS analysis of sulfur and phosphorus from the worn surfaces of the vee blocks.

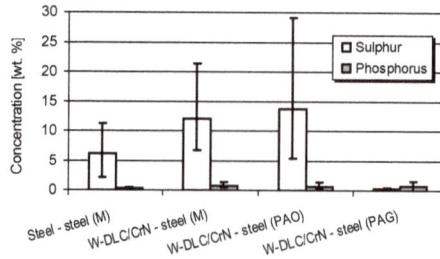

Figure 16. Quantitative results of the EDS analysis of sulfur and phosphorus from the worn surfaces of the pins.

As in the abrasion tests, the scatters of results showed a very inhomogeneous concentration of sulfur and phosphorus in the worn surface.

Under extreme pressure conditions during the scuffing tests, a dominating role was played by products of chemical reactions of sulfur [55], and organic compounds containing sulfur were more reactive under such conditions. Therefore, in almost all cases, the concentration of sulfur was much higher than phosphorus, unlike in the abrasion tests.

As in the abrasion tests, in the steel-steel tribosystem, in the worn area on the coated vee blocks, there were products of the reaction of sulfur with the steel surface—more than on the coating surface. In spite of this, the resistance to the scuffing of the coating-steel tribosystem was much higher than for steel-steel, irrespective of the oil used. As it was stated before, this was a result of the lower affinity of two different rubbing materials, than in the case of the uncoated contact, as well as the high hardness of the WC and CrN layers in the coating, leading to a smaller tendency for the creation of adhesive bonds and; therefore, there was a lower tendency for scuffing. Thus, as in the abrasion tests, the presence of the coating in the contact zone was of much greater significance than the chemical reactions with the oil.

Sulfur compounds found on the worn surface of the coated vee blocks were the result of the chemical reactions of EP additives in the oil with the steel material, which was transferred from the test pin onto the vee block, and the transfer of chemically modified steel from the pin. Such a transfer was also observed in the earlier work [15]. As also stated by other researchers (e.g., Mistry et al. [35] or Haque et al. [20]), the appearance of sulfur compounds on the worn surfaces of the test specimens adds beneficially to the improvement of the tribological features. The appearance of sulfur compounds in the friction zone, in the case of the coating-steel contact, adds to the prevention of scuffing.

For the PAG oil, the concentration of sulfur in the worn area of both test specimens was the lowest. This was the reason for the lowest resistance to scuffing, in this case, among the coating-steel tribosystems.

4.4. Pitting Tests

Figure 9, presented earlier, shows that, when the coating was deposited on the test cone, the resistance to pitting dropped significantly, irrespective of the oil used. It was also shown that the mineral and PAG oils exhibit similar resistance to pitting. The PAO oil gave a slightly worse result.

The reduction of the resistance to pitting was also observed by Fujii et al. [9] and by the authors of this paper [10,11] in case of the coated gears.

Results of SEM/EDS surface analysis from the worn surfaces of the test cones is shown in Figure 17.

As can be seen from Figure 17, irrespective of the oil used, the outer layers of the W-DLC/CrN coating were worn away, namely, the low-friction layer of WC/C and the hard, wear-resistant interlayer of WC (having a total thickness of 2.3 μm), exposing the adhesive layer of CrN. It was stated before that very big depths of the wear tracks on the coated cones were a result of plastic deformation of the substrate material. However, the presence of CrN layer is possible, because it is very ductile and adapts easily to the plastically deformed steel.

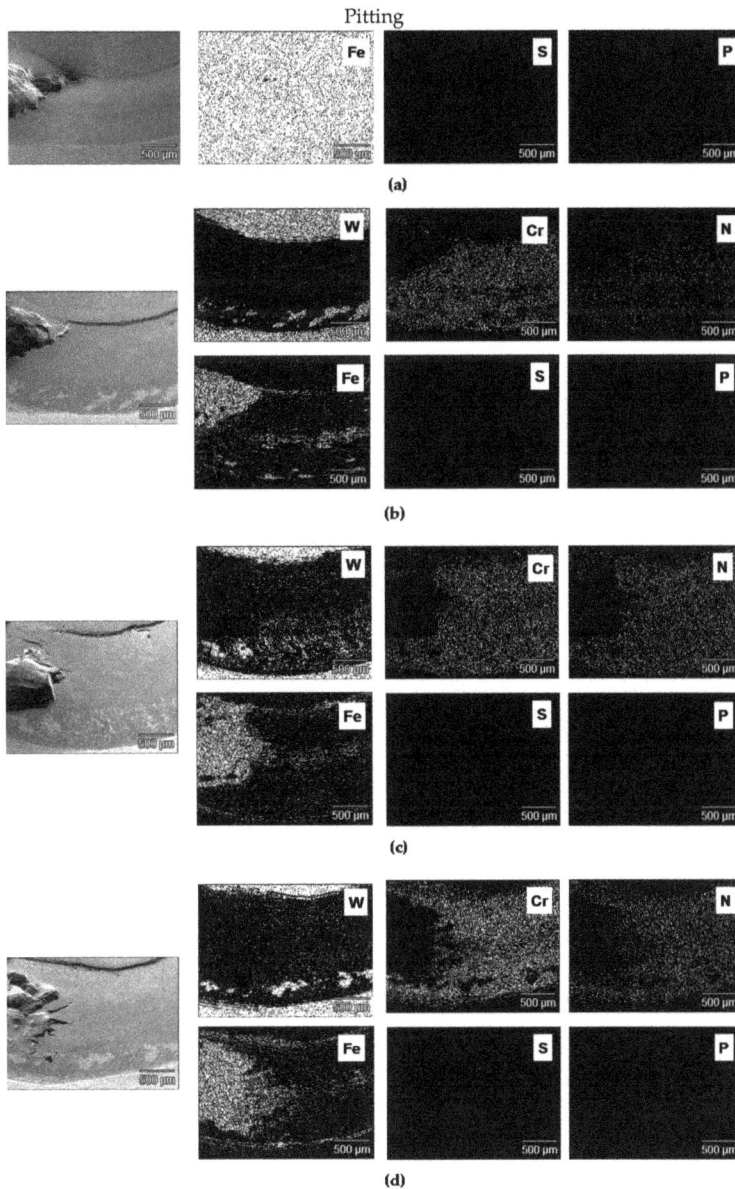

Figure 17. SEM images (on the left) and EDS surface analyses of the wear tracks on the test cones: (**a**) Uncoated cone lubricated with the mineral oil; (**b**) (W-DLC/CrN)-coated cone lubricated with the mineral oil; (**c**) (W-DLC/CrN)-coated cone lubricated with the PAO oil; and (**d**) (W-DLC/CrN)-coated cone lubricated with the PAG oil.

Results of quantitative EDS analysis of sulfur and phosphorus from the worn surfaces of the test cones are shown in Figure 18. Each analysis was performed three times at different points. The scatters between the minimum and maximum values were added to the graphs.

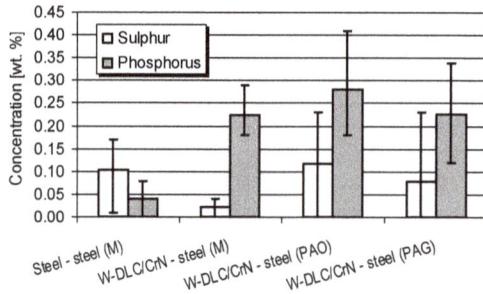

Figure 18. Quantitative results of the EDS analysis of sulfur and phosphorus from the worn surfaces of the test cones.

Unlike in the sliding contact, during the pitting tests, EP additives in the oil may have played an adverse role, because their high corrosion aggressiveness may lead to creation of numerous depressions and micropits on the lubricated surface, being potential nuclei for "macropits", reducing, in this way, the fatigue life. This was observed by, for example, Torrance et al. [56] and Tuszynski [57]. However, in the work by L'Hostis et al. [58], it was stated that a sulfur-rich film is formed at the tip of the crack. This film can act as both a barrier film towards hydrogen permeation within the metal and/or as an inhibitor of oil decomposition. The latter is associated with the nascent surface's ability to limit hydrogen generation. Without such hydrogen embrittlement, crack propagation is slowed down.

Mild friction conditions in the rolling contact during the pitting tests did not promote the chemical reactions of the EP additives in the oil with the steel surface. The concentration of sulfur and phosphorus in the surface layer was practically negligible. Thus, the better resistance to pitting given by the steel-steel tribosystem, than the coating-steel, irrespective of the oil used, can be explained from a physico-mechanical point of view.

The microhardness profiles in the surface layer were determined (Figure 19). The analyses were performed along three different lines. The confidence intervals representing 95% probability were calculated and included in the graphs.

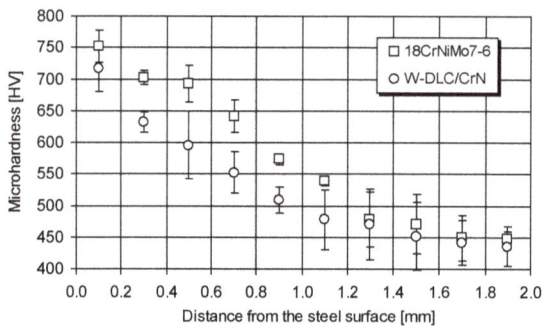

Figure 19. Microhardness profiles before and after deposition of the W-DLC/CrN coating.

From Figure 19, it was evident that, during the coating deposition process at the temperature of about 220–250 °C, due to phase transformations in the steel substrate, the microhardness dropped below the level observed for the uncoated steel. At the zone near the surface, which is most important for the fatigue life, the microhardness after deposition of W-DLC/CrN was lower than in the case of the uncoated sample. This correlates with the fatigue lives shown in Figure 9; the lower the near-surface microhardness, the lower the fatigue life.

Another possible reason for the reduction of the fatigue life by the coating is related to the coating surface (Figure 20).

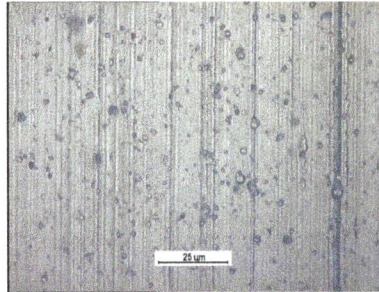

Figure 20. Optical microscope image of the surface of the W-DLC/CrN coating.

One could observe numerous droplets on the surfaces of the coating, which can be considered defects that initiate pitting.

Yet another reason for the decrease of the pitting resistance is related to the coating and substrate residual stress. Vackel and Sampath [59] found that coatings with higher compressive residual stress, hardness, and microstructural density enhanced the fatigue life of the coated specimens. However, coatings with tensile residual stress and somewhat weaker properties (hardness and stiffness) incurred a fatigue debit in the system. Varis et al. [60] observed that good fatigue performance was related to the compressive stress state of the substrate.

The phase transformations in the substrate due to coating deposition can cause a decrease in the residual stress, being an additional factor of the pitting acceleration in this research.

5. Conclusions

The final aim of the research was to improve the durability of the planetary transmissions in mining conveyors by the antifriction coating deposited on gear teeth.

The paper presents the results of the selection of the gear oil intended for gears made of case-hardened 18CrNiMo7-6 steel. The tests were performed using model, simple specimens.

The W-DLC/CrN antifriction coating was tested, representing an a-C:H:Me group. The coating was deposited on one of the specimens in the tribological system (e.g., on the vee blocks (abrasion and scuffing tests) and on the cones (pitting tests)), leaving the counter-specimens uncoated.

For lubrication, commercial industrial gear oils were used. They were the mineral oil, and two synthetic ones; one with a PAO base and one with a PAG base, and all with the viscosity grade of VG 320.

For the three gear oils lubricating the (W-DLC/CrN)-steel tribosystem, the resistance to abrasion notably rosein comparison with the steel-steel tribosystem lubricated with the mineral oil. The mineral and PAO oils gave the same results. PAG oil gave the worst result.

A decisive reason for the very good features of the coating-steel tribosystem in the abrasion tests is related to the action of the hard WC interlayer in the coating, and when WC is worn away, the CrN layer giving very good resistance to abrasion. The behavior of the oil depends on the chemical reactions of its EP additives with the surface of the test pin, especially. For the PAG oil, the concentration of sulfur in the worn area was the lowest. This is the reason for the lowest resistance to abrasion, in this case, among the coating-steel tribosystems.

For the three gear oils lubricating the (W-DLC/CrN)-steel tribosystem, the resistance to scuffing significantly rose in comparison with the steel-steel tribosystem lubricated with the mineral oil. The highest resistance to scuffing was given by the mineral and PAO oils. Similar to the behavior during the abrasion tests, PAG oil gave the worst result.

A decisive reason for the very good features of the coating-steel tribosystem in the scuffing tests is the lower affinity of two different rubbing materials, than in the case of the uncoated contact, as well as the high hardness of the WC and CrN layers in the coating, leading to a smaller tendency for the creation of adhesive bonds and; therefore, there was a lower tendency to scuffing.

As in the case of the abrasion tests, the behavior of the oil depends on the chemical reactions of its EP additives with the surface of the test pin, especially. For the PAG oil, the concentration of sulfur in the worn area was the lowest. This is the reason for the lowest resistance to scuffing, in this case, among the coating-steel tribosystems.

When the coating was deposited on one specimen, the resistance to pitting significantly dropped, irrespective of the oil used. The mineral and PAG oils exhibited similar resistance to pitting. The PAO oil gave a slightly worse result.

Unlike during the abrasion and scuffing tests performed in pure sliding contact, mild friction conditions in the rolling contact during the pitting tests did not promote the chemical reactions of the EP additives in the oil with the surface. Thus, the better resistance to pitting given by the steel-steel tribosystem than the coating-steel, irrespective of the oil used, can be explained from a physico-mechanical point of view. During the coating deposition process, due to phase transformations in the steel substrate, the microhardness dropped below the level observed for the uncoated steel. This correlates with the fatigue lives; the lower the near-surface microhardness, the lower the fatigue life.

To conclude, taking into consideration various criteria (i.e., the resistance to abrasion, scuffing, wear, and roughness of the test specimens), the PAO gear oil is well suited for the lubrication of the (W-DLC/CrN)-18CrNiMo7-6 tribosystem. Although this oil gave a slightly worse resistance to pitting than the other oils lubricating the coating–steel contact, the decisive reason for such a statement is that PAO exhibits a much better viscosity index than mineral oils, which is desired in the very extreme working conditions of mining conveyors. Such a material combination has a great application potential, since it gives the desired high resistance to abrasion and scuffing.

Author Contributions: Conceptualization, W.T. and R.M.; Methodology, W.T. and A.W.; Investigation, R.M., M.K., A.M.-S., E.O.-S., W.P., A.S.-A., and J.W.; Writing—Original Draft Preparation, W.T.; Writing—Review &Editing, M.S.; Visualization, W.T.; Supervision, R.M. and M.S.; and Funding Acquisition, A.W.

Funding: This research was founded by the National Centre for Research and Development (No. POIR.04.01.04-00-0064/15).

Conflicts of Interest: The authors declare no conflicts of interest.

Appendix A

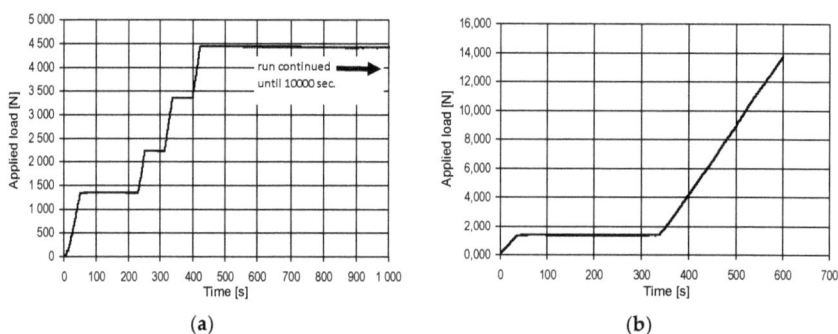

Figure A1. Load increase in (**a**) abrasion tests and (**b**) scuffing tests.

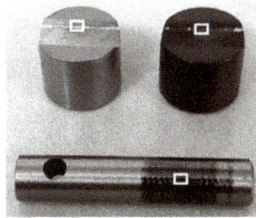

Figure A2. Tested vee blocks and pin: on the left—uncoated vee block; on the right—(W-DLC/CrN)-coated vee block.

Figure A3. Tested cones: on the left—uncoated cone; on the right—(W-DLC/CrN)-coated cone.

References

1. Tuszynski, W.; Kalbarczyk, M.; Michalak, M.; Michalczewski, R.; Wieczorek, A. The effect of WC/C coating on the wear of bevel gears used in coal mines. *Mater. Sci.* **2015**, *21*, 358–363. [CrossRef]
2. Kalin, M.; Vižintin, J. The tribological performance of DLC-coated gears lubricated with biodegradable oil in various pinion/gear material combinations. *Wear* **2005**, *259*, 1270–1280. [CrossRef]
3. Martins, R.C.; Moura, P.S.; Seabra, J.O. MoS$_2$/Ti low-friction coating for gears. *Tribol. Int.* **2006**, *39*, 1686–1697. [CrossRef]
4. Martins, R.; Amaro, R.; Seabra, J. Influence of low friction coatings on the scuffing load capacity and efficiency of gears. *Tribol. Int.* **2008**, *41*, 234–243. [CrossRef]
5. Michalczewski, R.; Kalbarczyk, M.; Tuszynski, W.; Szczerek, M. The scuffing resistance of WC/C coated spiral bevel gears. *Key Eng. Mater.* **2014**, *604*, 36–40. [CrossRef]
6. Ronkainen, H.; Elomaa, O.; Varjus, S.; Kilpi, L.; Jaatinen, T.; Koskinen, J. The influence of carbon based coatings and surface finish on the tribological performance in high-load contacts. *Tribol. Int.* **2016**, *96*, 402–409. [CrossRef]
7. Beilicke, R.; Bobach, L.; Bartel, D. Transient thermal elastohydrodynamic simulation of a DLC coated helical gear pair considering limiting shear stress behavior of the lubricant. *Tribol. Int.* **2016**, *97*, 136–150. [CrossRef]
8. Liu, H.; Zhu, C.; Zhang, Y.; Wang, Z.; Song, C. Tribological evaluation of a coated spur gear pair. *Tribol. Int.* **2016**, *99*, 117–126. [CrossRef]
9. Fujii, M.; Seki, M.; Yoshida, A. Surface durability of WC/C-coated case-hardened steel gear. *J. Mech. Sci. Technol.* **2010**, *24*, 103–106. [CrossRef]
10. Michalczewski, R.; Kalbarczyk, M.; Piekoszewski, W.; Szczerek, M.; Tuszynski, W. The rolling contact fatigue of WC/C-coated spur gears. *J. Eng. Tribol.* **2013**, *227*, 850–860. [CrossRef]
11. Michalczewski, R.; Piekoszewski, W.; Szczerek, M.; Tuszynski, W.; Antonov, M. The rolling contact fatigue of PVD coated spur gears. *Key Eng. Mater.* **2013**, *527*, 77–82. [CrossRef]
12. Benedetti, M.; Fontanari, V.; Torresani, E.; Girardi, C.; Giordanino, L. Investigation of lubricated rolling sliding behaviour of WC/C, WC/C-CrN, DLC based coatings and plasma nitriding of steel for possible use in worm gearing. *Wear* **2017**, *378*, 106–113. [CrossRef]
13. Singh, H.; Ramirez, G.; Eryilmaz, O.; Greco, A.; Doll, G.; Erdemir, A. Fatigue resistant carbon coatings for rolling/sliding contacts. *Tribol. Int.* **2016**, *98*, 172–178. [CrossRef]
14. Moorthy, V.; Shaw, B.A. Effect of as-ground surface and the BALINIT C and Nb–S coatings on contact fatigue damage in gears. *Tribol. Int.* **2012**, *51*, 61–70. [CrossRef]

15. Michalczewski, R.; Piekoszewski, W.; Szczerek, M.; Tuszynski, W. The lubricant-coating interaction in rolling and sliding contacts. *Tribol. Int.* **2009**, *42*, 554–560. [CrossRef]

16. Mercer, C.; Evans, A.G.; Yao, N.; Allameh, S.; Cooper, C.V. Material removal on lubricated steel gears with W-DLC-coated surfaces. *Surf. Coat. Technol.* **2003**, *173*, 122–129. [CrossRef]

17. Murakawa, M.; Komori, T.; Takeuchi, S.; Miyoshi, K. Performance of a rotating gear pair coated with an amorphous carbon film under a loss-of-lubrication condition. *Surf. Coat. Technol.* **1999**, *120*, 646–652. [CrossRef]

18. Topolovec-Miklozic, K.; Lockwood, F.; Spikes, H. Behaviour of boundary lubricating additives on DLC coatings. *Wear* **2008**, *265*, 1893–1901. [CrossRef]

19. Vengudusamy, B.; Mufti, R.A.; Lamb, G.D.; Green, J.H.; Spikes, H.A. Friction properties of DLC/DLC contacts in base oil. *Tribol. Int.* **2011**, *44*, 922–932. [CrossRef]

20. Haque, T.; Morina, A.; Neville, A.; Kapadia, R.; Arrowsmith, S. Effect of oil additives on the durability of hydrogenated DLC coating under boundary lubrication conditions. *Wear* **2009**, *266*, 147–157. [CrossRef]

21. Godfrey, D. Boundary lubrication. In *Interdisciplinary Approach to Friction and Wear*; Ku, P.M., Ed.; Southwest Research Institute: Washington, DC, USA, 1968; pp. 335–384.

22. Forbes, E.S. The load carrying action of organo-sulphur compounds—A review. *Wear* **1970**, *15*, 87–96. [CrossRef]

23. Coy, R.C.; Quinn, T.F.J. The use of physical methods of analysis to identify surface layers formed by organosulphur compounds in wear tests. *ASLE Trans.* **1975**, *18*, 163–174. [CrossRef]

24. Matveevsky, R.M.; Buyanovsky, I.A. *Antiseizure Properties of Lubricants under Conditions of Boundary Lubrication*; Izd. Nauka: Moscow, Russia, 1978. (In Russian)

25. Matveevsky, R.M.; Buyanovsky, I.A.; Karaulow, A.K.; Mischuk, O.A.; Nosovsky, O.I. Transition temperatures and tribochemistry of the surfaces under boundary lubrication. *Wear* **1990**, *136*, 135–139. [CrossRef]

26. Nakayama, K. Triboemission of charged particles from various solids under boundary lubrication conditions. *Wear* **1994**, *178*, 61–67. [CrossRef]

27. So, H.; Lin, Y.C. The theory of antiwear for ZDDP at elevated temperatures in boundary lubrication condition. *Wear* **1994**, *177*, 105–115. [CrossRef]

28. Johansson, E.; Hogmark, S.; Redelius, P. Surface analysis of lubricated sliding metal contacts. Part II. *Tribologia* **1997**, *16*, 26–38.

29. Piekoszewski, W.; Szczerek, M.; Tuszynski, W. The action of lubricants under extreme pressure conditions in a modified four-ball tester. *Wear* **2001**, *249*, 188–193. [CrossRef]

30. Tuszynski, W.; Piekoszewski, W. Effect of the type and concentration of lubricating additives on the antiwear and extreme pressure properties and rolling fatigue life of a four-ball tribosystem. *Lubr. Sci.* **2006**, *18*, 309–328. [CrossRef]

31. Kim, B.; Mourhatch, R.; Aswath, P.B. Properties of tribofilms formed with ashless dithiophosphate and zinc dialkyldithiophosphate under extreme pressure conditions. *Wear* **2010**, *268*, 579–591. [CrossRef]

32. Tuszynski, W.; Szczerek, M. Qualitative discrimination between API GL performance levels of manual transmission fluids by comparing their EP properties determined in a new four-ball scuffing test. *Tribol. Int.* **2013**, *65*, 57–73. [CrossRef]

33. Kalin, M.; Velkavrh, I.; Vižintin, J.; Ožbolt, L. Review of boundary lubrication mechanisms of DLC coatings used in mechanical applications. *Meccanica* **2008**, *43*, 623–637. [CrossRef]

34. Vlad, M.; Szczerek, M.M.; Michalczewski, R.; Kajdas, C.; Tomastik, C.; Osuch-Slomka, E. The influence of antiwear additive concentration on the tribological behaviour of a-C:H:W/steel tribosystem. *J. Eng. Tribol.* **2010**, *224*, 1079–1089. [CrossRef]

35. Mistry, K.K.; Morina, A.; Neville, A. A tribochemical evaluation of a WC-DLC coating in EP lubrication conditions. *Wear* **2011**, *271*, 1739–1744. [CrossRef]

36. Yue, W.; Liu, C.; Fu, Z.; Wang, C.; Huang, H.; Liu, J. Synergistic effects between sulfurized W-DLC coating and MoDTC lubricating additive for improvement of tribological performance. *Tribol. Int.* **2013**, *62*, 117–123. [CrossRef]

37. Yang, L.; Neville, A.; Brown, A.; Ransom, P.; Morina, A. Effects of lubricant additives on the WDLC coating structure when tested in boundary lubrication regime. *Tribol. Lett.* **2015**, *57*, 14. [CrossRef]

38. Yang, L.; Neville, A.; Brown, A.; Ransom, P.; Morina, A. Friction reduction mechanisms in boundary lubricated W-doped DLC coatings. *Tribol. Int.* **2014**, *70*, 26–33. [CrossRef]

39. Bobzin, K.; Brögelmann, T.; Stahl, K.; Stemplinger, J.P.; Mayer, J.; Hinterstoißer, M. Influence of wetting and thermophysical properties of diamond-like carbon coatings on the frictional behavior in automobile gearboxes under elasto-hydrodynamic lubrication. *Surf. Coat. Technol.* **2015**, *284*, 290–301. [CrossRef]

40. Forsberg, P.; Gustavsson, F.; Renman, V.; Hieke, A.; Jacobson, S. Performance of DLC coatings in heated commercial engine oils. *Wear* **2013**, *304*, 211–222. [CrossRef]

41. Al Mahmud, K.A.H.; Varman, M.; Kalam, M.A.; Masjuki, H.H.; Mobarak, H.M.; Zulkifli, N.W.M. Tribological characteristics of amorphous hydrogenated (a-C:H) andtetrahedral (ta-C) diamond-like carbon coating at different test temperatures in the presence of commercial lubricating oil. *Surf. Coat. Technol.* **2014**, *245*, 133–147. [CrossRef]

42. Xiao, Y.; Shi, W.; Luo, J.; Liao, Y. The tribological performance of TiN, WC/C and DLC coatings measured by the four-ball test. *Ceram. Int.* **2014**, *40*, 6919–6925. [CrossRef]

43. Lacey, P.I. Development of a gear oil scuff test (GOST) procedure to predict adhesive wear resistance of turbine engine lubricants. *Tribol. Trans.* **1998**, *41*, 307–316. [CrossRef]

44. Van de Velde, F.; Willen, P.; De Baets, P.; Van Geetruyen, C. Substitution of inexpensive bench tests for the FZG scuffing test—Part I: Calculations. *Tribol. Trans.* **1999**, *42*, 63–70. [CrossRef]

45. Van de Velde, F.; Willen, P.; De Baets, P.; Van Geetruyen, C. Substitution of inexpensive bench tests for the FZG scuffing test—Part II: Oil tests. *Tribol. Trans.* **1999**, *42*, 71–75. [CrossRef]

46. Bisht, R.P.S.; Singhal, S. A laboratory technique for the evaluation of automotive gear oils of API GL-4 level. *Tribotest* **1999**, *6*, 69–77. [CrossRef]

47. Trzos, M.; Szczerek, M.; Tuszynski, W. A study on the possibility of the Brugger test application for differentiation between the API-GL performance levels of gear oils. *Arch. Civ. Mech. Eng.* **2013**, *13*, 14–20. [CrossRef]

48. Polonsky, I.A.; Chang, T.P.; Keer, L.M.; Sproul, W.D. An analysis of the effect of hard coatings on near-surface rolling contact fatigue initiation induced by surface roughness. *Wear* **1997**, *208*, 204–219. [CrossRef]

49. Polonsky, I.A.; Chang, T.P.; Keer, L.M.; Sproul, W.D. A study of rolling-contact fatigue of bearing steel coated with physical vapor deposition TiN films: Coating response to cyclic contact stress and physical mechanisms underlying coating effect on the fatigue life. *Wear* **1998**, *215*, 191–204. [CrossRef]

50. Carvalho, N.J.M.; in't Veld, A.H.; De Hosson, J.T. Interfacial fatigue stress in PVD TiN coated tool steels under rolling contact fatigue conditions. *Surf. Coat. Technol.* **1998**, *105*, 109–116. [CrossRef]

51. Michalczewski, R.; Piekoszewski, W.; Szczerek, M.; Wulczyński, J. A method for the assessment of the rolling contact fatigue of modern engineering materials in lubricated contact. *Trans. FAMENA* **2012**, *36*, 39–48.

52. *ASTM D2625 Standard Test Method for Endurance (Wear) Life and Load-Carrying Capacity of Solid Film Lubricants (Falex Pin and Vee Method)*; ASTM International: West Conshohocken, PA, USA, 2015.

53. *ASTM D3233 Standard Test Methods for Measurement of Extreme Pressure Properties of Fluid Lubricants (Falex Pin and Vee Block Methods)*; ASTM International: West Conshohocken, PA, USA, 2014.

54. *IP300 Rolling Contact Fatigue Tests for Fluids in a Modified Four-Ball Machine*; Energy Institute (Formerly Institute of Petroleum): London, UK, 1982.

55. Stachowiak, G.; Batchelor, A.W. *Engineering Tribology*; Butterworth-Heinemann: Boston, MA, USA, 2001.

56. Torrance, A.A.; Morgan, J.E.; Wan, G.T.Y. An additive's influence on the pitting and wear of ball bearing steel. *Wear* **1996**, *192*, 66–73. [CrossRef]

57. Tuszynski, W. An effect of lubricating additives on tribochemical phenomena in a rolling steel four-ball contact. *Tribol. Lett.* **2006**, *24*, 207–215. [CrossRef]

58. L'Hostis, B.; Minfray, C.; Frégonèse, M.; Verdu, C.; Ter-Ovanessian, B.; Vacher, B.; Le Mogne, T.L.; Jarnias, F.; D'Ambros, A.D.-C. Influence of lubricant formulation on rolling contact fatigue of gears–interaction of lubricant additives with fatigue cracks. *Wear* **2017**, *382*, 113–122. [CrossRef]

59. Vackel, A.; Sampath, S. Fatigue behavior of thermal sprayed WC-CoCr-steel systems: Role of process and deposition parameters. *Surf. Coat. Technol.* **2017**, *315*, 408–416. [CrossRef]

60. Varis, T.; Suhonen, T.; Calonius, O.; Čuban, J.; Pietola, M. Optimization of HVOF Cr_3C_2NiCr coating for increased fatigue performance. *Surf. Coat. Techol.* **2016**, *305*, 123–131. [CrossRef]

coatings

MDPI

Article

Temperature-Induced Formation of Lubricous Oxides in Vanadium Containing Iron-Based Arc Sprayed Coatings

Wolfgang Tillmann [1], Leif Hagen [1],*, David Kokalj [1], Michael Paulus [2] and Metin Tolan [2]

[1]   Institute of Materials Engineering, TU Dortmund University, 44227 Dortmund, Germany;
      wolfgang.tillmann@tu-dortmund.de (W.T.); david.kokalj@tu-dortmund.de (D.K.)
[2]   Fakultät Physik/DELTA, TU Dortmund University, 44227 Dortmund, Germany;
      michael.paulus@tu-dortmund.de (M.P.); metin.tolan@tu-dortmund.de (M.T.)
*   Correspondence: leif.hagen@tu-dortmund.de; Tel.: +49-231-755-5715

Received: 12 November 2018; Accepted: 27 December 2018; Published: 29 December 2018

Abstract: In the field of surface engineering, the use of self-lubricous coatings with the incorporation of vanadium represent a promising approach to reduce friction, thus contributing to the wear behavior. For vanadium containing hard coatings produced by means of thin film technology, the reduction in friction at elevated temperatures was repeatedly attributed to temperature-induced and tribo-oxidatively formed oxides which act as solid lubricant. Only very few studies focused on the tribological characteristics of vanadium containing arc sprayed coatings. In this study, the tribological characteristics of a vanadium containing iron-based arc sprayed deposit were investigated in dry sliding experiments under ambient conditions and different temperatures. Types of wear at the worn surfaces and counterparts were examined by means of electron microscopy and energy dispersive X-ray (EDX) spectroscopy. The speciation of vanadium in the superficial layer was determined using X-ray absorption near edge structure (XANES) spectroscopy. It was found that the vanadium-containing coating exhibited a distinctly reduction of the coefficient of friction above 450 °C which further decreased with increasing temperature. XANES spectroscopy indicated an increased oxidation state for the V component on the coating surface, suggesting the prevalence of specific vanadium oxides which promote a self-lubricating ability of the coating.

Keywords: wire arc spray; friction behavior; lubricous oxides; XANES spectroscopy

1. Introduction

Tribo-oxidation was validated for practical applications as a favorite mechanism for the development of dry running tribosystems [1]. Over the last decades, self-lubricious hard coatings evolved into a promising candidate to provide an improved friction behavior, which in turn contributes to the wear resistance [2]. In the field of thin film technology, a reduction in friction has been achieved, among others, by using V-containing coatings [3]. For V-containing coatings, temperature-induced and tribo-oxidatively formed, low shear strength oxides act as solid lubricants, thus determining the lubricating capacity of functional surfaces. The participation of V forms oxides of the homologous series V_nO_{2n-1} with $3 < n < 10$ between the end members V_2O and VO_2, and are so-called Magnéli phases [4,5]. In addition, V_nO_{2n+1} type phases exist, where VO_2 and V_2O_5 represent the end members [5]. Some V-oxides belong to another homology according to the formula $V_{2n}O_{5n-2}$ [6] and are named as Wadsley phases. In summary, a number of compounds in the V–O system can be formed including stoichiometric and substoichiometric oxides, where V is present in number of valence states [7]. The slip ability of oxygen deficient phases of V-oxides has been univocally associated as crystallographic shear [8]. When compared to intrinsic solid lubricants such as MoS_2, where every second layer features a crystallographic slip ability, only few layers depending on the stoichiometry exhibits a crystallographic shear structure in Magnéli phases. Woydt et al. [9] therefore clarified that a univocal association between the lubricity and Magnéli phases is not always correct. Regarding the lubricity of Magnéli phases, they emphasized that the rheological process can be better described by plastic flow instead of crystallographic shear.

V-oxides or their derivatives can be deposited, on the one hand, as coatings using reactive physical vapor deposition techniques [10–12], hydrothermal synthesis [13], and atmospheric pressure chemical vapor deposition [14]. On the other hand, an established approach is to deposit hard and wear-resistant coatings with the incorporation of V. Thereby, V acts as an active element source for the formation for various oxides that are primarily caused by tribo-chemical reactions at elevated temperatures in ambient air. In that respect, a major objective has been focused on the development of hard coatings with the participation of V incorporated as a solid solution [15–18]. Further studies emphasized the technological relevance of multilayered coating systems consisting of alternating thin layers of VN and another hard coating [19–23]. Concerning the tribological characteristics, the aforementioned studies revealed that the reduce in friction was attributed to certain types of V-oxides. As recently discussed [24], the temperature-induced formation of V-oxides depends, inter alia, on the V content and chemical composition of the coating system. Due to its low decohesion energy [5], vanadium pentoxide represents a promising candidate for solid lubricants at elevated temperatures. Above its melting point of 678 °C [25], it also acts as liquid lubricant [2]. Kutschej et al. [26] pointed out that the transformation of vanadium pentoxide into lower oxidized V-oxides with higher melting points in turn limits the lubricating capacity.

To date, only few authors examined the tribo-mechanical properties of arc sprayed coatings using a V-containing feedstock. Deng et al. [27] investigated the abrasive wear behavior of arc sprayed stainless steel coatings which contain a certain amount of V. The incorporation of V was accomplished by the addition of ferrovanadium as filler material. The authors demonstrated an enhanced wear resistance against abrasion, among others, due to the addition of V, which was found to favor the formation of vanadium carbides. With regard to the tribological investigations, the friction behavior of the V-containing coating was not investigated at all.

In this study, an arc-sprayed V-containing iron based coating is subjected to elevated temperatures and tested tribologically in dry sliding experiments. To scrutinize the wear mechanism of the tribological stressed surfaces, the wear tracks and counterbodies are investigated by means of electron microscopy and EDX spectroscopy. To characterize the temperature-induced oxide formation, the oxidation state of the V component of the reactive layer at the surface is determined by using XANES spectroscopy.

2. Materials and Methods

Prior to the coating deposition, round steel specimens (1.1191, 40 mm × 6 mm) were grit blasted with corundum (grit size: 1180–1700 μm), and cleaned in an ultrasonic ethanol bath. A Fe-based cored wire which contains approximately 29.9 wt % of V served as feedstock. Within this study, the produced coating which was deposited with the V-containing Fe-based (Fe-29.91V-0.38Al-0.33Si-0.25Mn-0.07C wt %) cored wire was referred to as Fe-V. To investigate the effect of V on the temperature-induced and tribo-oxidatively promoted oxide formation, a corresponding V-free Fe-based coating was produced with the use of a low alloy steel in the form of a cored wire (Fe-0.25Mn-0.11Si-0.07C), and consequently served as reference. The chemical composition of the cored wires was provided by the manufacture (Durum Verschleissschutz, Willich, Germany), and calculated based on the results conducted by chemical analysis of the steel strip (Fe-based sheath) and optical emission spectrometry of the filler material (ferrovanadium, low-alloy steel powder).

To produce the arc sprayed coatings, the Smart Arc 350 PPG spraying system (Oerlikon Metco, Pfäffikon, Switzerland) with the spray torch equipped with a conventional front-end hardware (high-profile centering post, No. PPG5I976; air cap body (fine), No. PPG51416 [28]) was utilized.

The optimized spray parameter settings were used based on the findings in a previous study [29] using a statistical design of experiments and multi-criteria optimization by means of Derringer's desirability function. This approach has been taken in order to produce dense and smooth coatings with an almost defect-free microstructure. In that respect, compressed air was utilized as atomization gas by applying a gas pressure of 0.6 MPa. The voltage and current was set to 28 V, and 180 A. The deposits were applied using a constant spray distance of 95 mm. The gun velocity of the spray torch and the track pitch were set to 200 mm/s, and 5 mm. Two overruns were executed in order to generate an adequate coating thickness of 418 ± 34 μm.

To ensure functional integrity, the produced deposits were metallographically examined. Cross sections of coated samples were prepared by using silicon carbide grinding discs (grit size: P80, P180; P600, and P2500 according to FEPA [30]) and polishing cloths with a diamond suspension (abrasive particle size: 9, 6, 3 and 1 μm). The microstructural characteristics of the produced coatings were examined at cross sections using a field emission scanning electron microscope (FE-SEM) type JSM-7001F (JEOL (Germany) GmbH, Freising, Germany). EDX spectra were acquired with an energy dispersive X-ray spectroscopy detector (Oxford Instruments, Abingdon, UK) in conjunction with the FE-SEM. To determine the chemical composition, the EDX spectra were evaluated with the use of the INCA software (Oxford Instruments, Abingdon, UK).

Coated samples from the same batch were further investigated with respect to the tribo-mechanical properties. Prior to the tribo-mechanical testing, the surface of the coated samples was machined using a silicon carbide grinding disk (grit size: P4000), and polished utilizing polishing cloths with diamond suspension (abrasive particle size: 3 μm and $1/4$ μm). The tribological tests were performed using the high-temperature ball-on-disk (BOD) tribometer (CSM Instruments, Peseux, Switzerland) at room temperature (25 °C) and elevated temperatures (from 350 up to 750 °C in 100 °C steps). The experiments were conducted without lubricant supply. For the dry sliding experiments, an alumina ball with a diameter of 6 mm and a hardness of 2300 HV0.3 was used as counterbody. The velocity was kept constant at 40 cm/s, and the radius of the circular path was set to 10 mm. A sliding distance of 200 m, and a load of 5 N were applied. The coefficient of friction (COF) was obtained from the relationship of the measured tangential force and the applied normal force. The fluctuations of the tangential force during sliding were extracted and afterwards analyzed by Fast Fourier Transformation (FFT) using the data analysis and graphing software OriginPro (OriginLab Corporation, Northampton, MA, USA). In order to assess the wear behaviour of the tribological stressed surfaces, the worn surfaces as well as the alumina counterbodies were investigated by means of electron microscopy and EDX spectroscopy, as described previously. Vickers microhardness (MH) measurements were conducted on the coating surfaces after tribological testing using the hardness testing device M400 (Leco, Saint Joseph, MI, USA) with 300 g loads. To determine the MH, five residual indents were taken into account.

Due to the rapid solidification (about ~10^5 °C/s) [31] of molten spray particles, arc spraying has great potential for the production of amorphous coatings, as already demonstrated in [32,33] for Fe-based coatings. XANES spectroscopy is a well-established technique for examination of the chemical structure of amorphous materials. It offers a unique possibility to identify the local structural environment of the atomic species in a material of unknown composition. Some XANES features are sensitive to the local symmetry and oxidation state of the exciting atom. Within the scope of this study, the oxidation state of the reactive layer at the surface of the Fe-V coating was analyzed using XANES at beamline BL10 of the synchrotron light source DELTA (Dortmund electron accelerator) at the TU Dortmund. The XANES measurements were conducted at the V K-edge (5465 eV). The measuring range was set to 60 eV before the V K-edge (5465 eV), and up to 100 eV after the K-edge. A step size of 2 eV was chosen before the K-edge, whereas the measurement was performed using a step size of 0.5 eV in the region of the K-edge in order to get an adequate resolution. After the K-edge the step size was set to 1 eV. A measurement time of 5 s per step was used. A silicon (111) channel-cut crystal served as a monochromator. The beam was set to 1 mm in height and 4 mm in width. The angle of incidence was 5.0°, and the detection of the absorption was carried out by fluorescence. Different V-oxides (MaTeck, Jülich, Germany) were utilized for the XANES spectroscopy and served as reference for further examinations of the Fe-V coating. A V-foil (space group Im-3m, body-centered cubic structure) was used as calibration data. The measurements were conducted ex-situ at the surface of the coatings which have been subjected to the tribological tests under elevated temperatures. The obtained raw data (XANES spectra) was extracted using the Software PyMca [34] of the European Synchrotron Radiation Facility. To examine the features of the XANES spectra, the data was normalized via standard edge-step normalization, i.e., reduced by background subtraction with a linear function. The absolute energy zero point was chosen in relation to the first inflection point of the V metal derivative spectrum (called E_0), which corresponds to the convection of the excitation of an inner shell electron to an empty state just above the Fermi edge of the V metal. The V foil was scanned to correct the energy shift and to obtain energy-calibrated spectra in a consistent manner. To investigate the specific XANES features, the program package Athena (Demeter Bruce Ravel's XAS Data Analysis Software [35]) was employed. In essence, the absorption edge was taken as the energy of the first point of inflection of the principal absorption edge (given by the maximum in the derivative spectrum), which can be assigned to the energy half way up the normalized-edge step, i.e., where the absorption is equal to 0.5 ($E_{1/2}$). In terms of the pre-edge feature, the data several electronvolts before and after the pre-edge feature was extracted separately, and then fitted with a pseudo-Voigt function as demonstrated in [36]. Afterwards, the pre-edge feature was scrutinized by calculating the pre-edge peak centroid position, pre-edge peak area, and pre-edge peak intensity using the data analysis and graphing software OriginPro as described above.

3. Results and Discussion

3.1. Tribological Investigation

Prior to the tribological examination, the samples used were initially investigated by means of electron microscopy in order to ensure functional integrity. SEM images showing the cross section confirm that the coating consists of a dense and almost defect-free microstructure with no delamination or cracks (see Supplementary Materials, Figure S1). With regard to the chemical composition, EDX analyses confirm that the produced Fe-V coating consists of 55.6 ± 1.2 wt % of Fe, 26.4 ± 0.5 wt % of V, and 15.2 ± 1.0 wt % of O (other bal. wt %). In this respect, the Fe-V coating is interstratified with oxides between individual lamellae. Macroscopically, the arc sprayed Fe-V coating possesses a lamellar microstructure, whereby the individual lamellae exhibit a slight different chemical composition with varying thicknesses. In terms of the oxide formation, XRD analysis [29] verified that the Fe-V coating is mainly composed of V_2O_3, VO and Fe_3O_4. Nevertheless, it is stated that not all phases were identified, since phase transformation processes result in a wide varying non-equilibrium state due to constitutional supercooling during rapid solidification. In addition, the crystallographic data of many metastable phases are not present in the databases.

Figure 1 shows the COF of the arc sprayed Fe-V coating which was obtained from the tribological testing at various temperatures. The V-free steel coating serves as reference. The COF of the Fe-V coating slightly but constantly falls between 25 and 550 °C from 0.66 ± 0.15 to 0.55 ± 0.09. A significant decrease of the COF is observed between 550 and 650 °C (0.55 ± 0.09 → 0.36 ± 0.07). When the sample is exposed to a temperature of 750 °C, the COF further decreases to 0.29 ± 0.05. Opposed to that, the V-free reference exhibits an almost constant COF up to 750 °C of approximately 0.50 ± 0.07. The findings indicate a distinct reduced COF for the Fe-V coating above 550 °C, especially at 750 °C. At 750 °C and under same environmental conditions, the Fe-V coating exhibits a friction reduction of approximately 42% in contrast to the V-free reference. At the same time, the COF decreases by 56% when compared to 25 °C.

(a)

(b)

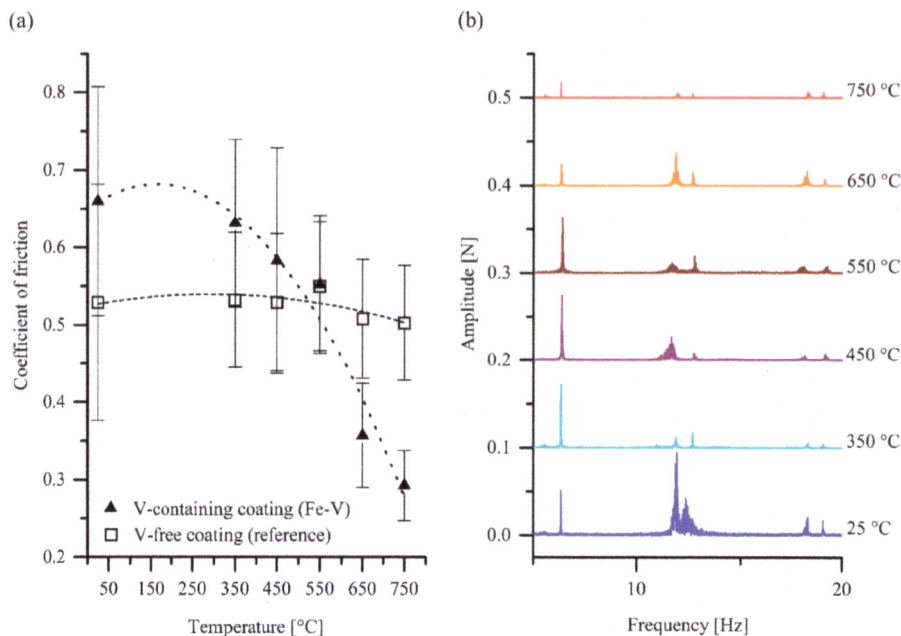

Figure 1. (**a**) Coefficient of friction in dependency on the operating temperature, and (**b**) frequency of the amplitude spectrum calculated from the extracted data of the tangential force measurement during sliding in dependency on the operating temperature.

Stacked FFT analyses of the measured amplitude spectrum revealed that no additional frequencies occurred when the tribological stressed surfaces were subjected to different operating temperatures (Figure 1b). The amplitude expresses the oscillation of the friction forces around the steady state as described in [37]. The spectrum of the amplitude of the coatings is the same for all analyzed temperatures. However, it is striking that the magnitude of the amplitude observed for different temperatures differs distinctly. For instance, at 25 °C the amplitude of the tangential force reaches the highest value, whereas the measurement of the surface which was tribological stressed at 750 °C shows the lowest value. Between the aforementioned temperatures, the amplitude of the friction processes reveals no continuous change with increasing temperature. The findings encourage the assumption that the tribological stressed surface progresses a more constant sliding at elevated temperature, in particular at 750 °C. Reduction of the amplitude is also reported for $CrV(x)N$ thin films for the increase of the temperature from 25 to 550 °C.

The worn surfaces were examined after the tribological tests by means of electron microscopy and EDX spectroscopy (Figure 2). Since the backscattered electrons (BSE) signal is related to the atomic number, an amendment in chemical composition can be obtained. As seen from the SEM images (Figure 2a), mild oxidational regimes are observed across the tribological stressed contact area which was tested at ambient air without any heat treatment. As indicated by electron microscopy, the oxides are mainly formed at cavities, and also in areas of surface disruption which results in certain breakouts of particles. At elevated temperatures up to 450 °C, the Fe-V coating exhibits a tribo-oxidatively formed thin film which can be seen across the wear track. The BSE signal confirms a different chemical composition between the wear track and the surrounding surface area. EDX analyses (Figure 2b) verify an increased amount of oxides across the wear track, when compared to the surrounded surface area. Starting with 550 °C, SEM images show the appearance of oxide grains covering a large part of the surface. At 650 and 750 °C, the surfaces are prone to form various mixed oxides consisting of different proportions of Fe and V covering the entire surface. As obtained from the BSE signal, the wear track and enclosed surface area exhibit a similar chemical composition. A similar distribution of elements is additionally confirmed by EDX spot analyses. For the tribological stressed surface at 750 °C, EDX analyses reveal that both the surface oxidation and tribo-oxidation is amplified. Thus, it can be concluded that above 450 °C the amount of oxides on the coating surface is significantly higher compared to 25 °C (Figure 2b).

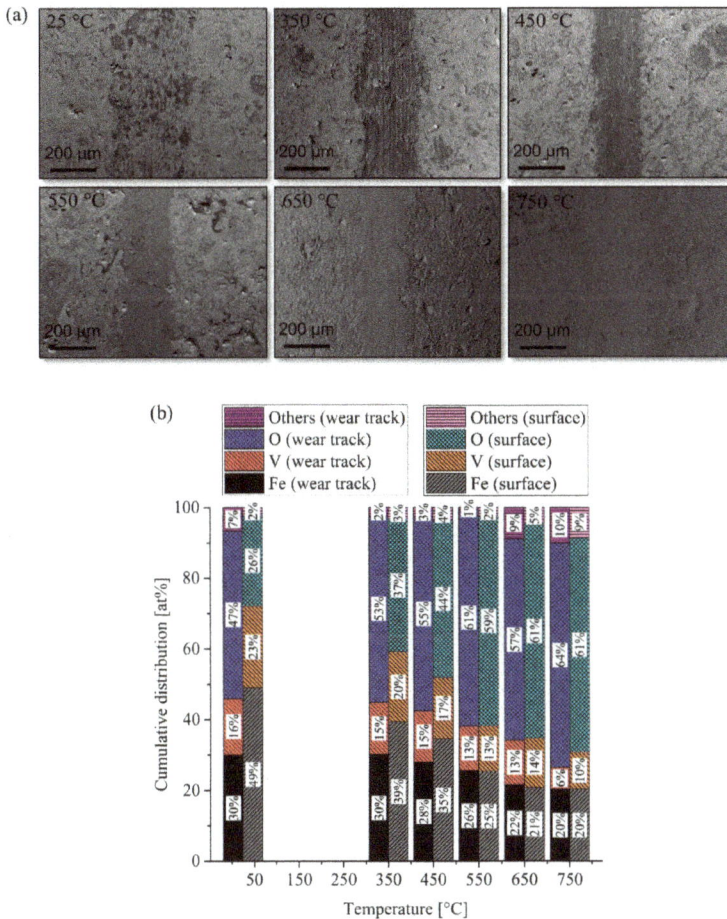

Figure 2. (a) SEM images showing the wear tracks after tribological testing, and (b) distribution of elements determined by EDX in at % at different spots in dependency of the operating temperature during tribological testing.

Figure 3 shows a magnified view of the wear tracks after sliding at 550, 650 and 750 °C as well as the corresponding SEM images which were detected by the BSE and secondary electrons (SE) signal. With regard to the tribological stressed surface which has been exposed to a temperature of 550 °C during testing, it can be seen that the contact zone predominantly exhibits a covering oxide film (Figure 3, BSE-mode at 550 °C). Individual areas in turn show a local surface disruption (Figure 3, SE-mode at 550 °C).

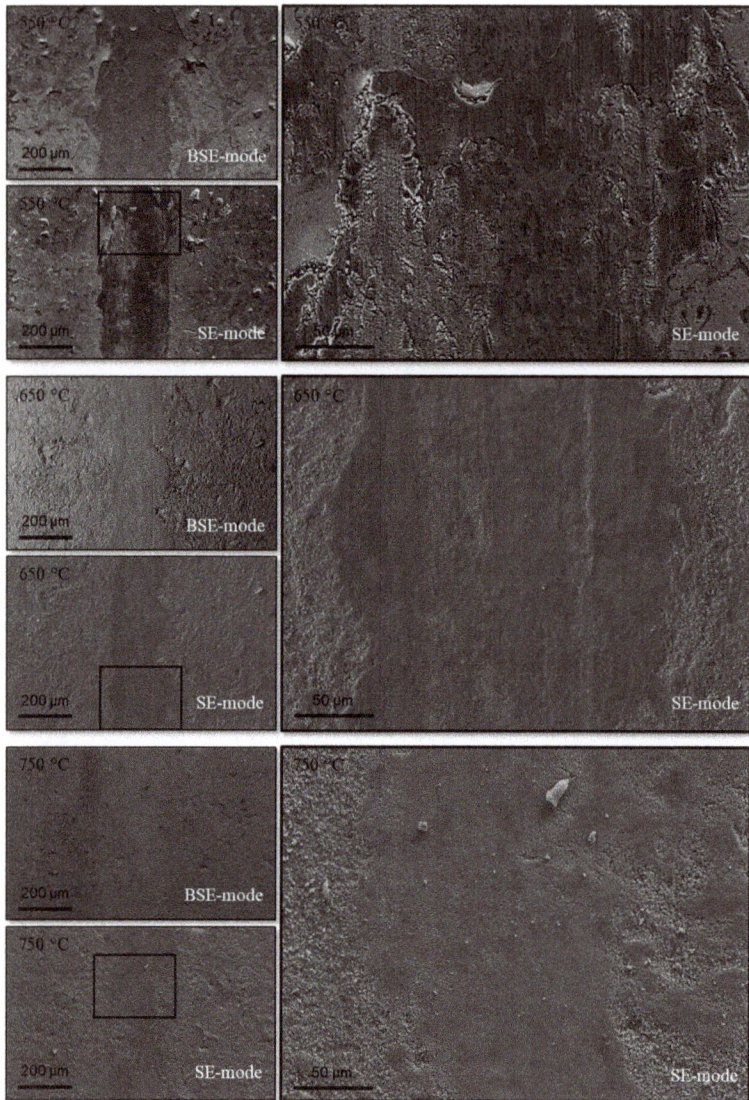

Figure 3. SEM images showing the wear tracks after tribological testing at 550, 650 and 750 °C, using different contrasting methods as well as in a magnified view.

A number of small grooves in the direction of sliding are visible in a magnified view. It is stated the torn out wear particles have been sliding between the two sliding surfaces, and thus, dragging across the softer Fe-V coating. As a result, the wear particles which are ploughing through the Fe-V coating remove the material and cause abrasion such as grooves, indicating three-body wear [38].

For the tribological stressed surface at 650 °C, the material is smeared in the direction of sliding (Figure 3, SE-mode at 650 °C). Shear tongues are formed and a slight amount is drawn out in the direction of sliding. At 750 °C, the coating surface exhibits no significant material removal. In contrast, it can be seen that oxide grains are smoothened and compacted (Figure 3, SE-mode at 750 °C).

A so-called "glaze" is formed at the contact area which was running against the alumina counterbody. It is assumed that the glaze is formed from sintered wear particles such as oxide grains.

Figure 4 shows the wear flat on the alumina counterbody after sliding against the coated sample at 550, 650, and 750 °C. As shown in the SEM images, the alumina counterbodies are hardly abraded. Substantially, material accumulation at the wear flat on the alumina counterbody shows an adhesive wear. However, the counterbodies examined at different temperatures show a different extent of wear. Concerning the examinations above 25 °C, it is found that with increasing temperature a larger amount of material adheres onto the wear flat of the alumina counterbody. Such adhesion is more pronounced at 550 °C when compared to 25, 350 or 450 °C (data not shown). Above 550 °C (650 °C → 750 °C), a distinctly increased amount of material adheres onto the wear flat of the alumina counterbody. In addition, that material is strongly smeared and elongated in the sliding direction. A comparison between the different counterbodies shows that for the counterbody, which was sliding against the coating at 750 °C, the smeared material is much more elongated in the sliding direction. For the adhered material observed at various temperatures, EDX analyses confirm that the material is composed of V-(Fe-) rich oxides. However, as demonstrated in Figure 4, EDX spot analyses (see Spectrum 1, 3 and 5) verify an increased V content in the adhered material, in particular after the dry sliding experiments at 650 and 750 °C. Since it can be assumed that the alumina counterbody is chemically inert, the Al content in the region of the adhered material (Spectrum 1, 3 and 5) can be explained by the inherent characteristics of the EDX spectroscopy, i.e., the interaction between the primary electron beam and the alumina (EDX spot analyses were carried out with 20 keV). According to the findings, it can be concluded that a certain amount of the material originated from the tribological stressed surface adheres on the alumina counterbody.

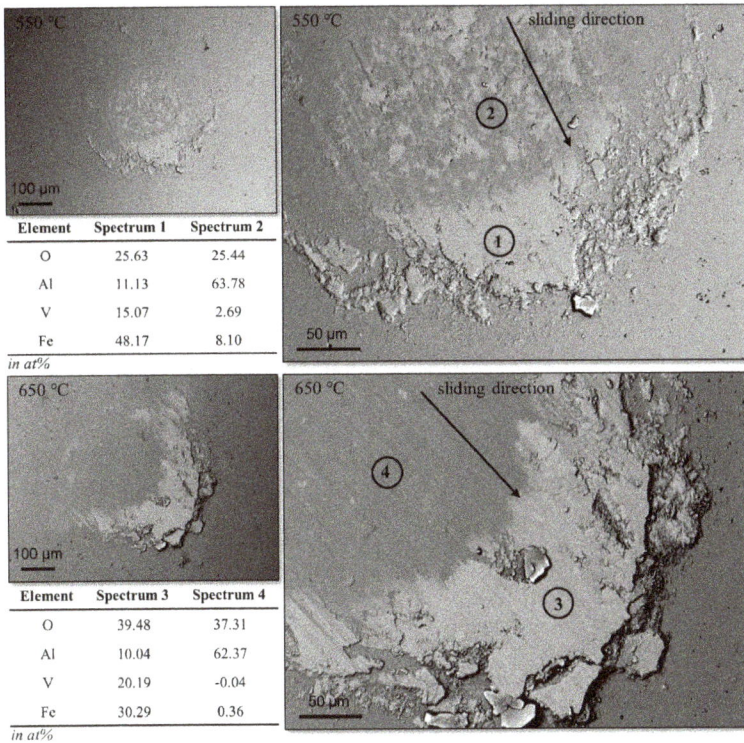

Element	Spectrum 1	Spectrum 2
O	25.63	25.44
Al	11.13	63.78
V	15.07	2.69
Fe	48.17	8.10

in at%

Element	Spectrum 3	Spectrum 4
O	39.48	37.31
Al	10.04	62.37
V	20.19	-0.04
Fe	30.29	0.36

in at%

Figure 4. *Cont.*

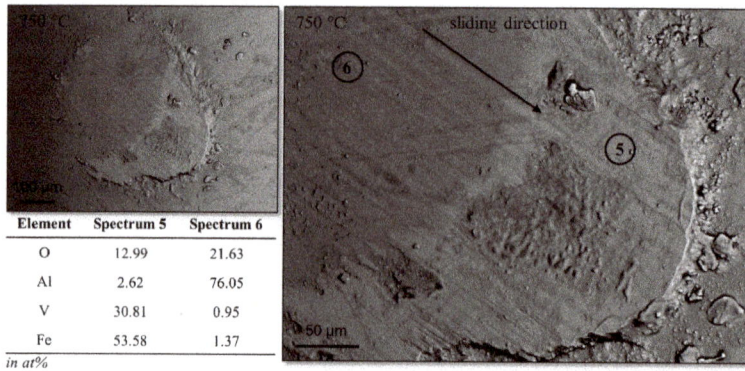

Figure 4. SEM images showing the wear flat on the alumina counterbody after sliding against the coated sample depending on the operating temperature. For the marked regions (Spectrums 1 to 6), the chemical composition is determined by means of EDX.

3.2. XANES Spectroscopy

In order to constrain the average V oxidation state of the coatings, XANES spectra at the V K-edge of several references (vanadium oxides, metallic vanadium), which corresponds to different valence states, were initially measured. The phase purity for V_2O_5 and V_2O_3 was confirmed by XRD. The measurements indicate that the V_2O_5 (V^{5+}) features an orthorhombic crystalline structure, whereas the V_2O_3 (V^{3+}) exists in a trigonal crystalline structure. Further analyses of the other reference oxides confirm no phase purity. VO (V^{2+}) exists as the cubic phase, and includes also the monoclinic V_2O_4 phase. V_2O_4 (V^{4+}) consists mainly of the tetragonal phase, whereby it also includes some amount of the orthorhombic and monoclinic phase as well as orthorhombic V_2O_5. The V_6O_{13} ($V^{4.3+}$) reference features a monoclinic crystalline structure with some traces of V_2O_3 and V_2O_5. Normalized XANES spectra for the references investigated in this study, i.e., various V oxides, are shown in Figure 5a.

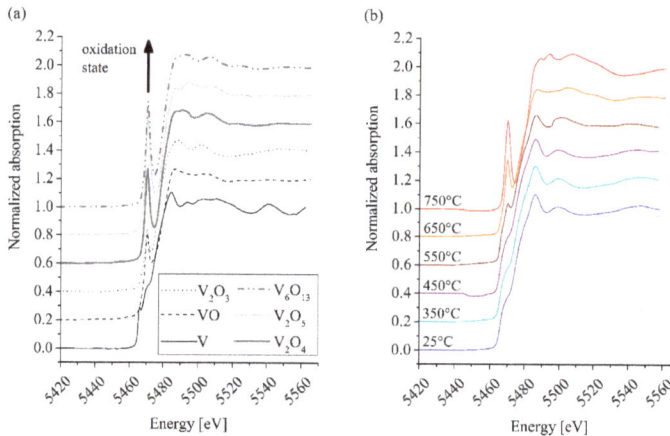

Figure 5. XANES spectra at the V K-edge of (**a**) several vanadium oxides and metallic vanadium, as well as (**b**) XANES spectra at the V K-edge of the coating surface for various temperatures.

Evaluating the different XANES spectra, it is found that the spectra demonstrate significant variations in certain features such as the energy position and intensity of the peaks in the pre-edge region as well as in the edge region. With respect to the absorption edge position, it is shown the edge energy $E_{1/2}$ of the references (Figure 5a, Table 1) demonstrates a shift to higher energies with increase in the oxidation state of V. For instance, the energy position of the absorption edge tends to rise from 8.7 eV after the K-edge for V^{2+} to 13.8 eV for V^{5+}. The energy shifts are assigned to Kunzl's law as shown in [39], which states that as the valence of the central absorbing atom (i.e., absorbing V atom) increases, all absorption features shift towards higher energy. It needs to be mentioned that the energy shift follows Kunzl's law when a comparison is made among compounds with the same metal-ligand combinations. However, the unambiguous determination of the valence states with the help of the Kunzl's law for some cases is difficult, since the ranges of the chemical shift for two different valences can sometimes overlap. Chaurand et al. [40] pointed out that shape of the XANES spectra at the absorption edge is quite sensitive to the valence state. However, they also reported that the features at the absorption edge are susceptible to the local atomic surrounding, or to a range of interferences such as multiscattering effects arising from more distant neighbors around the central V atoms. Several vanadium oxides, showing an oxidation state from 0 to 5, were analyzed by means of XANES spectroscopy. As obtained from the normalized XANES spectra, the pre-edge peak features of references show distinct variations as well (Figure 5a, Table 1). The V-foil exhibits a pre-edge peak at a centroid position of 5466 eV. Additionally, a subsequent shoulder at 5471 eV is observed before the absorption increases. Opposed to that, the V oxides are characterized by a pre-edge peak centroid position at 5470.5 eV indicating an amendment of the oxidation state or symmetry. The intensity of the pre-edge maximum increases from 0.29 for V to 0.85 for V_2O_5, which corresponds with the chemical alteration (i.e., oxidation state) of V, as shown in Table 1.

Table 1. Characteristic XANES features of the vanadium oxides and metallic vanadium.

Phase	Oxidation-State	Pre-Edge		Main-Edge Position (eV)
		Normalized Intensity	Position (eV)	
V	0	0.29	1	9.5
VO	2	0.62	5.5	8.7
V_2O_3	3	0.44	5.5	10.8
V_2O_4	4	0.67	5.5	13.2
V_6O_{13}	4.3	0.75	5.5	14.4
V_2O_5	5	0.85	5.5	13.8

According to Rees et al. [41], the pre-edge feature is related to electronic transitions from the 1s core levels to the empty 3d levels. As discussed in [42], the transition is induced due to the mixing of 3d orbitals of vanadium with 2p orbitals of oxygen. Studying the speciation of V in oxide phases from steel slag, Chaurand et al. [40] concluded that V^{3+} compounds could be assigned to a pre-edge peak intensity of 0.05–0.10, whereas V^{4+} compounds and V^{5+} compounds could be linked to an intensity of 0.30–0.65 and 0.50–1.20, respectively, depending on the local symmetry. In this respect, V-oxides which consist of a tetrahedral or pyramidal symmetry were characterized by a more intense pre-edge maximum when compared to an octahedral symmetry [40,43]. The authors claimed that the intensity gets influenced, inter alia, by diverse energy resolution by using different monochromator reflections. As a result, the pre-edge peak intensity allows an insufficient discrimination of the vanadium valence. Quoting their discussion on different methodologies examining the pre-edge features [39,44–49], Chaurand et al. [40] in turn clarified that the correlation between the pre-edge peak centroid position and its integrated area represents the most reliable method for determining the speciation of V. Thus, substantial changes occur in both the energy position (i.e., centroid position) and total area of the pre-edge peak.

The normalized XANES spectra measured at the not worn surface of the arc sprayed Fe-V coatings after tribological testing at different temperatures are illustrated in Figure 5b. The characteristic features obtained from the XANES spectra (i.e., normalized pre-edge maximum intensity, centroid position, and main edge position) are summarized in Table 2.

Table 2. Characteristic XANES features of the arc sprayed V-containing Fe-based coating after tribological testing at different temperatures.

Coating Temperature (°C)	Pre-Edge		Main-Edge Position (eV)
	Normalized Intensity	Position (eV)	
25	0.39 *	5 *	9.5
350	0.39 *	5 *	9.5
450	0.43 *	5 *	9.5
550	0.47	5	9.8
650	0.57	5	11.6
750	0.74	5	12.5

The absorption at 25, 350 and 450 °C runs a similar course and shows no pronounced pre-edge peak (marked with * in Table 2), which in turn indicate no distinct oxidation. Starting from 550 °C, a pre-edge peak is gradually formed, that is more intense at 650 and 750 °C. Furthermore, for the same temperature increase (550 °C → 750 °C), the absorption edge shifts to higher energies suggesting an onset of oxidation. In terms of the near-edge structure above the onset of the absorption edge, the different form of oscillation indicates the emergence of new phases with higher oxidation states and different local symmetry. As verified by Farges [50], those features at the absorption edge are predominantly sensitive to the valence and local atomic surrounding of the absorbing element. However, as reported by the same author, the features are also influenced by single or multiple scattering effects as well as changes in the medium and long range environment.

In the next step, we investigate the correlation between the normalized pre-edge peak area and the pre-edge centroid position of the treated Fe-V coatings as well as of the vanadium oxides (Figure 6).

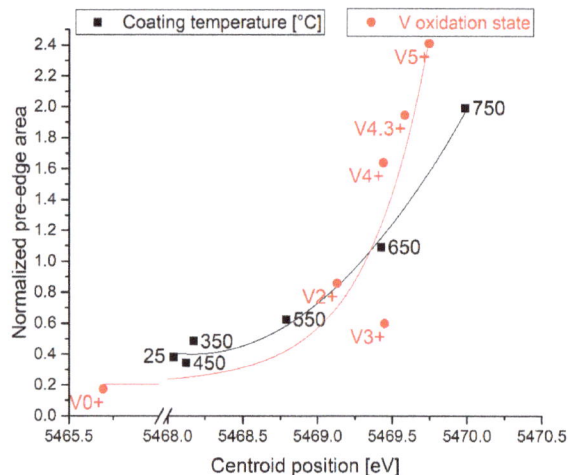

Figure 6. Dependence of pre-edge maximum centroid position and pre-edge maximum area of different treated Fe-V coatings and vanadium oxides (references).

It is found that both the pre-edge peak centroid position and the normalized area of the corresponding peak increase with a larger oxidation state of V. Whereas the normalized pre-edge peak area of pure V is small, a strong increase occurs with rising oxidation state. Since the centroid position and normalized area of the pre-edge peak depend on the crystal structure, the irregularity between V^{2+} and V^{3+} can be attributed to the impurity of references owing to a mixture of different vanadium oxides. For a temperature of 25 °C, the Fe-V coating reveals a pre-edge peak centroid position of 5468.0 eV and a normalized area of 0.38, which corresponds to an oxidation state higher than V^{0+} and lower than V^{2+}, or V^{3+} respectively. XRD analysis of the as-sprayed coating reveals the occurrence of V_2O_3 and VO as reported in a previous study [29]. Since the oxidation state of VO and V_2O_3 is 2 respectively 3, the position of the pre-edge peak obtained by the XANES spectra illustrates that the coating is not fully oxidized and contains un-oxidized vanadium. Accordingly, the Fe-V coating in its initially state (i.e., as-sprayed and polished conditions) is already interstratified with V-rich oxides which is due to the use of compressed air as atomization gas as well as the environmental conditions during spraying in ambient air. A heat treatment of the Fe-V coating at 350 and 450 °C shows no considerable influence on the pre-edge peak characteristics. Correspondingly, no change in the phase composition is obtained by means of XRD. Hence, the heat treatment at 350 and 450 °C in ambient air does not influence the oxidation state of the V component as measured at the coating surface. The slight increase in surface oxidation up to 450 °C (see EDX, Figure 2b) may be traced back to the formation of Fe-rich oxides. This in turn corresponds to the BSE signal as demonstrated in Figure 2a, indicating some amplified tribo-oxidation phenomena at the wear track with the participation of Fe-rich oxides. In contrast, an increased temperature of 550 °C reveals an amendment of the oxidation state between V^{2+} and V^{3+} as shown by a pre-edge peak centroid position of 5468.8 eV with a normalized pre-edge peak area of 0.62. According to the XRD analyses, due to the heat treatment at 550, 650 and 750 °C the formation of VO_2 and V_2O_5 as well as ternary oxides of the Fe-V-O system takes place [29]. However, the analysis of the XRD patterns is not distinct since a superposition of the reflections of miscellaneous oxides is present. Accordingly, XANES spectroscopy can be used to estimate the average oxidation state of the coatings at different temperatures. For a temperature of 650 °C, a pre-edge peak centroid position of 5469.4 eV and a normalized pre-edge peak area of 1.09 are calculated. Compared to the reference oxides, the determined pre-edge peak characteristics reveal an oxidation state between V^{3+} and V^{4+}. As verified by the SEM images, the coating surface is subjected to a significant oxidation. A further increase in temperature up to 750 °C causes an ongoing oxidation process to an oxidation state between $V^{4.3+}$ and V^{5+} as shown by a pre-edge peak centroid position of 5470.0 eV and normalized pre-edge peak area of 1.99. Comparing both the fit of the reference oxides depending on the oxidation state and the fit of the heat treated Fe-V coating at different temperatures (Figure 6), it is to state that the general pathway of the two fits differs. In fact, the references (Figure 5a) consist of compounds of the binary V–O system, whereas the formation of miscellaneous oxides related to the Fe–V–O system or other phases of the Fe–V system on the coating surface (Figure 5b) could not be ruled out. Accordingly, the scattering of the radiation on other atoms, such as Fe, can affect the absorption and consequently the position and shape of the peaks [51]. However, it can be concluded that the oxidation state of the V component, as detected on the surface of the Fe–V coating, increases with increasing temperature. Similar findings were observed studying the oxidation behavior of AlCrV$_x$N thin films at various temperatures [24]. With respect to the tribological investigations (Figure 1), it is found that the COF initially drops gradually until a temperature of 450 °C, which then decreases significantly with increasing temperature starting from 550 °C. Since the oxidation state of the V component increases above 450 °C, it is believed that the oxidation state of the V component in the reactive layer has a fundamental impact on the resulting friction behavior (i.e., the alumina counterbody has a high chemical inertness). Quoting the discussion on various studies on V-containing hard coatings, the authors recently clarified in [24] that the reduction of the COF is attributed, among others, to the formation of certain Magnéli phases. Despite a ferrovanadium alloy as feedstock, the oxidation of vanadium is fostered at elevated temperature, leading to a possible formation of self-lubricous vanadium oxides which causes the reduction of the COF. The pre-oxidized state of the Fe-V coating under as-sprayed conditions does not influence the ongoing oxidation process at elevated temperatures.

4. Conclusions

A vanadium-containing iron-based coating was deposited by means of arc spraying with a resulting vanadium content of up to 26.4 ± 0.5 wt %. As verified in dry sliding experiments, the COF of the vanadium-containing coating decreases gradually until reaching a temperature of 450 °C. With further increases in temperature starting from 550 °C the COF drops significantly. In contrast, a vanadium-free reference features an almost constant COF up to 750 °C. Evaluating the dry sliding behavior under same environmental conditions, i.e., at a temperature of 750 °C, the vanadium-containing coating possesses a friction reduction of approximately 42% compared to the vanadium free reference. Simultaneously, the COF decreases by 56% when compared to 25 °C. As confirmed by certain pre-edge peak characteristics using XANES spectroscopy, the reduction of friction starting from 550 °C corresponds with an average oxidation state higher than V2+. Further surface oxidation phenomena at 650 and 750 °C, respectively, are attributed to a further increased oxidation state for the V component.

Supplementary Materials: The following are available online at http://www.mdpi.com/2079-6412/9/1/18/s1, Figure S1: SEM image showing the cross section of the produced V-containing Fe-based coating.

Author Contributions: Conceptualization, L.H.; Investigation, L.H., D.K. and M.P.; Writing-Original Draft Preparation, L.H., D.K. and M.P.; Writing-Review & Editing, L.H. and D.K.; Supervision, W.T. and M.T.; Project Administration, L.H.

Funding: This research received no external funding.

Acknowledgments: The contributions of DURUM Verschleissschutz GmbH are gratefully acknowledged for their support in providing the vanadium-containing feedstock. The authors would like to thank the DELTA machine group for providing synchrotron radiation.

Conflicts of Interest: The authors declare no conflict of interest.

References

1. Woydt, M.; Skopp, A.; Dörfel, I.; Witke, K. Wear Engineering Oxides/Antiwear Oxides©. *Tribol. Trans.* **1999**, *42*, 21–31. [CrossRef]
2. Franz, R.; Mitterer, C. Vanadium containing self-adaptive low-friction hard coatings for high-temperature applications: A review. *Surf. Coat. Technol.* **2013**, *228*, 1–13. [CrossRef]
3. Brugnara, R.H. Hochtemperaturaktive HPPMS-Verschleißschutzschichten durch Bildung reibmindernder Magnéli-Phasen im System (Cr,Al,X)N. Ph.D. Thesis, Technische Hochschule Aachen, Aachen, Germany, January 2016. (In German)
4. Stegemann, B.; Klemm, M.; Horn, S.; Woydt, M. Switching adhesion forces bycrossing the metal-insulator transition in Magneli-type vanadium oxide crystals. *Beilstein J. Nanotechnol.* **2011**, *2*, 59–65. [CrossRef] [PubMed]
5. Reeswinkel, T.; Music, D.; Schneider, J.M. Ab initio calculations of the structure and mechanical properties of vanadium oxides. *J. Phys. Condens. Matter* **2009**, *21*, 145404. [PubMed]
6. Lamsal, C.; Ravindra, N.M. Optical properties of vanadium oxides-an analysis. *J. Mater. Sci.* **2013**, *48*, 6341–6351. [CrossRef]
7. Hryha, E.; Rutqvist, E.; Nyborg, L. Stoichiometric vanadium oxides studied by XPS. *Surf. Interface Anal.* **2012**, *44*, 1022–1025. [CrossRef]
8. Kharton, V. *Solid State Electrochemistry I: Fundamentals, Materials and their Applications*; John Wiley & Sons: Weinheim, Germany, 2009.
9. Woydt, M.; Skopp, A.; Dörfel, I.; Witke, K. Wear engineering oxides/anti-wear oxides. *Wear* **1998**, *218*, 84–95.
10. Lugscheider, E.; Bärwulf, S.; Barimani, C. Properties of tungsten and vanadium oxides deposited by MSIP-PVD process for self-lubricating applications. *Surf. Coat. Technol.* **1999**, *120–121*, 458–464. [CrossRef]
11. Lugscheider, E.; Knotek, O.; Bobzin, K.; Bärwulf, S. Tribological properties, phase generation and high temperature phase stability of tungsten- and vanadium-oxides deposited by reactive MSIP-PVD process for innovative lubrication applications. *Surf. Coat. Technol.* **2000**, *133–134*, 362–368. [CrossRef]
12. Gulbiński, W.; Suszko, T.; Sienicki, W.; Warcholiński, B. Tribological properties of silver- and copper-doped transition metal oxide coatings. *Wear* **2003**, *254*, 129–135. [CrossRef]
13. Vernardou, D.; Louloudakis, D.; Spanakis, E.; Katsarakis, N.; Koudoumas, E. Electrochemical properties of vanadium oxide coatings grown by hydrothermal synthesis on FTO substrates. *New J. Chem.* **2014**, *38*, 1959–1964. [CrossRef]

14. Louloudakis, D.; Vernardou, D.; Spanakis, E.; Katsarakis, N.; Koudoumas, E. Electrochemical properties of vanadium oxide coatings grown by APCVD on glass substrates. *Surf. Coat. Technol.* **2013**, *230*, 186–189. [CrossRef]

15. Qui, Y.; Zhang, S.; Lee, J.W.; Li, B.; Wang, Y.; Zhao, D.; Sun, D. Towards hard yet self-lubricious CrAlSiN coatings. *J. Alloy. Comp.* **2015**, *618*, 132–138.

16. Fernandes, F.; Loureiro, A.; Polcar, T.; Cavaleiro, A. The effect of increasing V content on the structure, mechanical properties and oxidation resistance of Ti-Si-V-N films deposited by DC reactive magnetron sputtering. *Appl. Sur. Sci.* **2014**, *289*, 114–123. [CrossRef]

17. Bobzin, K.; Bagcivan, N.; Ewering, M.; Brugnara, R.H.; Theiss, S. DC-MSIP/HPPMS (Cr,Al,V)N and (Cr,Al,W)N thin films for high-temperature friction reduction. *Surf. Coat. Technol.* **2011**, *205*, 2887–2892. [CrossRef]

18. Franz, R.; Neidhardt, J.; Mitterer, C.; Schaffer, B.; Hutter, H.; Kaindl, R.; Sartory, B.; Tessadri, R.; Lechthaler, M.; Polcik, P. Oxidation and diffusion processes during annealing of AlCrVN hard coatings. *J. Vac. Sci. Technol. A* **2008**, *26*, 302–308. [CrossRef]

19. Chang, Y.Y.; Chiu, W.T.; Hung, J.P. Mechanical properties and high temperature oxidation of CrAlSiN/TiVN hard coatings synthesized by cathodic arc evaporation. *Surf. Coat. Technol.* **2016**, *303 Pt A*, 18–24. [CrossRef]

20. Qiu, Y.; Zhang, S.; Lee, J.W.; Li, B.; Wang, Y.; Zhao, D. Self-lubricating CrAlN/VN multilayer coatings at room temperature. *Appl. Surf. Sci.* **2013**, *279*, 189–196. [CrossRef]

21. Luo, Q. Temperature dependent friction and wear of magnetron sputtered coating TiAlN/VN. *Wear* **2011**, *271*, 2058–2066. [CrossRef]

22. Qui, Y.; Zhang, S.; Li, B.; Wang, Y.; Lee, J.W.; Li, F.; Zhao, D. Improvement of tribological performance of CrN coating via multilayering with VN. *Surf. Coat. Technol.* **2013**, *231*, 357–363.

23. Park, J.K.; Baik, Y.J. Increase of hardness and oxidation resistance of VN coating by nanoscale multilayered structurization with AlN. *Mater. Lett.* **2008**, *62*, 2528–2530. [CrossRef]

24. Tillmann, W.; Kokalj, D.; Stangier, D.; Paulus, M.; Sternemann, C.; Tolan, M. Investigation on the oxidation behavior of AlCrV$_x$N thin films by means of synchrotron radiation and influence on the high temperature friction. *Appl. Surf. Sci.* **2018**, *427*, 511–521. [CrossRef]

25. Wriedt, H.A. The O–V (Oxygen-Vanadium) system. *Bull. Alloy Phase Diagr.* **1989**, *10*, 271–277. [CrossRef]

26. Kutschej, K.; Mayrhofer, P.H.; Kathrein, M.; Polcik, P.; Mitterer, C. A new low-friction concept for Ti$_{1-x}$Al$_x$N based coatings in high-temperature applications. *Surf. Coat. Technol.* **2004**, *188–189*, 358–363. [CrossRef]

27. Deng, Y.; Yu, S.F.; Yan, N.; Xing, S.L.; Huang, L.B. Effect of vanadium and niobium on abrasive behaviour of arc sprayed 4Cr13 coatings. *Appl. Mech. Mater.* **2013**, *395*, 712–717. [CrossRef]

28. PL 40933 EN 07 PPG SmartArc®Gun Parts List (EN). Available online: https://www.oerlikon.com/metco/en/products-services/coating-equipment/thermal-spray/spray-guns/spray-guns-arc/ppg/ (accessed on 27 December 2018).

29. Tillmann, W.; Hagen, L.; Kokalj, D.; Paulus, M.; Tolan, M. A study on the tribological behavior of vanadium-doped arc sprayed coatings. *J. Therm. Spray Technol.* **2017**, *26*, 503–516. [CrossRef]

30. Federation of European Producers of Abrasives. FEPA Grains Standards. Available online: https://www.fepa-abrasives.com/abrasive-products/grains (accessed on 27 December 2018).

31. Newbery, A.P.; Grant, P.S.; Neiser, R.A. The velocity and temperature of steel droplets during electric arc spraying. *Surf. Coat. Technol.* **2005**, *195*, 91–101. [CrossRef]

32. Guo, W.; Wu, Y.; Zhang, J.; Yuan, W. Effect of the long-term heat treatment on the cyclic oxidation behavior of Fe-based amorphous/nanocrystalline coatings prepared by high-velocity arc spray process. *Surf. Coat. Technol.* **2016**, *307*, 392–398. [CrossRef]

33. Cheng, J.; Zhao, S.; Liu, D.; Feng, Y.; Liang, X. Microstructure and fracture toughness of the FePSiB-based amorphous/nanocrystalline coatings. *Mater. Sci. Eng. A* **2017**, *696*, 341–347. [CrossRef]

34. Solé, V.A.; Papillon, E.; Cotte, M.; Walter, P.; Susini, J. A multiplatform code for the analysis of energy-dispersive X-ray fluorescence spectra. *Spectrochim. Acta B* **2007**, *62*, 63–68. [CrossRef]

35. Ravel, B.; Newville, M. ATHENA, ARTEMIS, HEPHAESTUS: Data analysis for X-ray absorption spectroscopy using IFEFFIT. *J. Synchrotron Radiat.* **2005**, *12*, 537–541. [CrossRef] [PubMed]

36. Poumellec, B.; Marucco, J.F.; Touzelin, B. X-ray-absorption near-edge structure of titanium and vanadium in (titanium,vanadium) dioxide rutile solid solutions. *Phys. Rev. B* **1987**, *35*, 2284–2294. [CrossRef]

37. Perfilyev, V.; Moshkovich, A.; Lapsker, I.; Laikhtman, A.; Rapoport, L. The effect of vanadium content and temperature on stick–slip phenomena under friction of CrV(x)N coatings. *Wear* **2013**, *307*, 44–51. [CrossRef]

38. Hutchings, I.; Shipway, P. Wear by hard particles. In *Tribology*, 2nd ed.; Hutchings, I., Shipway, P., Eds.; Butterworth-Heinemann: Oxford, UK, 2017; pp. 165–236.

39. Wong, J.; Lytle, F.W.; Messmer, R.P.; Maylotte, D.H. K−edge absorption spectra of selected vanadium compounds. *Phys. Rev. B* **1984**, *30*, 5596–5610. [CrossRef]

40. Chaurand, P.; Rose, J.; Briois, V.; Salome, M.; Proux, O.; Nassif, V.; Olivi, L.; Susini, J.; Hazemann, J.L.; Bottero, J.Y. New methodological approach for the vanadium K-edge X-ray absorption near-edge structure interpretation: Application to the speciation of vanadium in oxide phases from steel slag. *J. Phys. Chem. B* **2007**, *111*, 5101–5110. [CrossRef] [PubMed]

41. Rees, J.A.; Wandzilak, A.; Maganas, D.; Wurster, N.I.C.; Hugenbruch, S.; Kowalska, J.K.; Pollock, C.J.; Lima, F.A.; Finkelstein, K.D.; DeBeer, S. Experimental and theoretical correlations between vanadium K-edge X-ray absorption and Kβ emission spectra. *J. Biol. Inorg. Chem.* **2016**, *21*, 793–805. [CrossRef]

42. Tanaka, T.; Yamashita, H.; Tsuchitani, R.; Funabiki, T.; Yoshida, S. X-ray absorption (EXAFS/XANES) study of supported vanadium oxide catalysts. Structure of surface vanadium oxide species on silica and (γ-alumina at a low level of vanadium loading. *J. Chem. Soc. Faraday Trans. 1* **1988**, *84*, 2987–2999. [CrossRef]

43. Krause, B.; Darma, S.; Kaufholz, M.; Mangold, S.; Doyle, S.; Ulrich, S.; Leiste, H.; Stuber, M.; Baumbach, T. Composition-dependent structure of polycrystalline magnetron-sputtered V-Al-C-N hard coatings studied by XRD, XPS XANES and EXAFS. *J. Appl. Crystallogr.* **2013**, *46*, 1064–1075. [CrossRef]

44. Giuli, G.; Paris, E.; Mungall, J.; Romano, C.; Dingwell, D. V oxidation state and coordination number in silicate glasses by XAS. *Am. Mineral.* **2004**, *89*, 1640–1646. [CrossRef]

45. Sutton, S.R.; Karner, J.; Papike, J.; Delaney, J.S.; Shearer, C.; Newville, M.; Eng, P.; Rivers, M.; Dyar, M.D. Vanadium K edge XANES of synthetic and natural basaltic glasses and application to microscale oxygen barometry. *Geochim. Cosmochim. Acta* **2005**, *69*, 2333–2348. [CrossRef]

46. Passerini, S.; Smyrl, W.H.; Berrettoni, M.; Tossici, R.; Rosolen, M.; Marassi, R.; Decker, F. XAS and electrochemical characterization of lithium intercalated V_2O_5 xerogels. *Solid State Ion.* **1996**, *90*, 5–14. [CrossRef]

47. Wilke, M.; Farges, F.; Petit, P.E.; Brown, G.E.J.; Martin, F. Oxidation state and coordination of Fe in minerals: An Fe K-XANES spectroscopic study. *Am. Mineral.* **2001**, *86*, 714–730. [CrossRef]

48. Wilke, M.; Partzsch, G.M.; Bernhardt, R.; Lattard, D. Determination of the iron oxidation state in basaltic glasses using XANES at the K-edge. *Chem. Geol.* **2004**, *213*, 71–87. [CrossRef]

49. Petit, P.-E.; Farges, F.; Wilke, M.; Solé, V.A. Determination of the iron oxidation state in Earth materials using XANES pre-edge information. *J. Synchrotron Radiat.* **2001**, *8*, 952–954. [CrossRef]

50. Farges, F. *Ab initio* and experimental pre-edge investigations of the Mn K-edge XANES in oxide-type materials. *Phys. Rev. B* **2005**, *71*, 155109. [CrossRef]

51. Rehr, J.J.; Albers, R.C. Theoretical approaches to x-ray absorption fine structure. *Rev. Mod. Phys.* **2000**, *72*, 621–654. [CrossRef]

coatings

Article

Effect of TiO$_2$ Sol and PTFE Emulsion on Properties of Cu–Sn Antiwear and Friction Reduction Coatings

Lixia Ying [1,*], Zhen Fu [1], Ke Wu [1], Chunxi Wu [1], Tengfei Zhu [1], Yue Xie [1] and Guixiang Wang [2]

[1] College of Mechanical and Electrical Engineering, Harbin Engineering University, Harbin 150001, China; fuzhen0826@163.com (Z.F.); kirkwuke@163.com (K.W.); wuchunxi789789@163.com (C.W.); zhutengf89115@163.com (T.Z.); xieyue@hrbeu.edu.cn (Y.X.)

[2] College of Materials Science and Chemical Engineering, Harbin Engineering University, Harbin 150001, China; wangguixiang@hrbeu.edu.cn

* Correspondence: yinglixia@hrbeu.edu.cn

Received: 22 November 2018; Accepted: 16 January 2019; Published: 19 January 2019

Abstract: The aim of this paper is to obtain Cu–Sn composite coatings incorporated with PTFE and TiO$_2$ particles, which have superior antiwear and friction reduction properties. Electrodeposition was carried out in a pyrophosphate electrolyte, and the electrochemical behavior of the plating solutions was estimated. PTFE emulsion and TiO$_2$ sol were prepared and used, of which the average particle sizes were less than 283 and 158 nm, respectively. Then, four different types of coatings, Cu–Sn, Cu–Sn–TiO$_2$, Cu–Sn–PTFE and Cu–Sn–PTFE–TiO$_2$, were electroplated with a pulsed power supply. Their microstructure, composition, microhardness, corrosion resistance and tribological properties were then analyzed and compared in detail. The results show that both PTFE and TiO$_2$ are able to improve coating structure and corrosion resistance, while they have different effects on hardness and tribological properties. However, the presence of both PTFE and TiO$_2$ in the deposited coating leads to a lower friction coefficient of 0.1 and higher wear and corrosion resistance.

Keywords: electrodeposition; Cu–Sn; PTFE; TiO$_2$ sol; tribological properties

1. Introduction

Copper–Tin (Cu–Sn) alloys are widely applied to various kinds of friction parts for their excellent self-lubricating properties, which can effectively reduce friction and wear under oil-free lubrication conditions [1,2]. In recent decades, electroplating Cu–Sn base coatings has attracted extensive interest. Initially, researchers focused on experimental parameters and electrolyte composition regarding enhanced properties. Various baths appeared and were used for electrodeposition of Cu–Sn alloy, such as phosphate fluoborate, boron–fluoride, pyrophosphate and cyanide based [3,4]. In addition, some new process methods were explored to enhance the self-lubricating properties of Cu–Sn composite coatings [5].

In recent years, adding self-lubricating particles to plating solutions seems to have become an effective method to further reduce the friction coefficient of coatings. It is reported that the applying of multi-walled carbon nanotubes reduces the friction coefficient and wear loss of Copper-Tin alloy to 28% and 32%, respectively [6]. That aside, as a potential lubricating material, polytetrafluoroethylene (PTFE) is usually adapted to modify compositing coatings. With the addition of PTFE to plating solutions, Balaji et al. [7] and Du et al. [8] obtained Cu–Sn–PTFE and Cu–Sn–Zn–PTFE composite coatings, respectively. Both of the coatings possessed superior self-lubricating properties. In the course of friction, the composite particles precipitated from the coating and formed a lubricating film to reduce friction and wear [9].

Evidently, the presence of PTFE soft particles lowers the friction coefficient of the Cu–Sn alloy but the shortcomings of a Cu–Sn alloy are also obvious, including softness and weak carrying capacity.

More importantly, the coatings are seriously damaged under a heavy load. Therefore, it becomes more and more important to improve the hardness and wear resistance of the composite coatings and meet the requirements of a heavier load and harsh working conditions.

There are some studies with the aim of improving the composite coatings' hardness and wear resistance by co-depositing hard particles or through various electrodeposition methods [10–13]. With the addition of nano-Al_2O_3 to the electroless Ni–P–PTFE alloy plating solution, Xu et al. [14] developed a Ni–P–PTFE–Al_2O_3 composite coating with increased hardness and wear resistance. Chen et al. [15] obtained a Ni–P matrix composite coating containing nano-Al_2O_3 particles and PTFE particles by utilizing nano-Al_2O_3 and PTFE in the plating solution. The evidence shows that the incorporation of two different kinds of particles enhances the wear resistance along with reducing the friction coefficient. In fact, nano-sized particles are more prone to agglomerating in an electroplating bath. Although various surfactants were adapted in an attempt to disperse nano-particles, it was still difficult to obtain a uniformly dispersed solution, which is one of the main problems associated with fabrication of nano-composite coatings.

To overcome the non-uniformity of the dispersion particles, coating preparation was also conducted through combining a sol method with a traditional electrodeposition method by some researchers. Chen et al. added TiO_2 Sol into the conventional acid Ni electroplating solution to strengthen the coatings [16,17]. The research showed that the nano TiO_2 particles embedded in the deposited metal matrix restrain the growth of the deposited metal, leading to the formation of a more uniform and fine microstructure. Wang et al. [18,19] obtained sol-reinforced Ni–B–TiO_2 and Au–Ni–TiO_2 nano-composite coatings. Compared with Ni–B and Au–Ni coatings, respectively, their mechanical properties and wear resistance improved greatly along with the increasing nano-hardness. Based on those analyses, it can be concluded that sol is a highly dispersed system which can replace traditional powder in the plating solution to promote uniform dispersion of hard particles in the coating. Once nano-TiO_2 is embedded in the deposited metal layer, and restrain the growth of the deposited metal, a more uniform and fine microstructure could be achieved.

Thus, the objective of the present study is to obtain a Cu–Sn base composite coating with superior comprehensive properties. Based on this, TiO_2 sol and PTFE emulsion were prepared and added into the solution to codeposite with the Cu–Sn alloy. Simultaneously, the electrodepositional behavior of the bath, the particle size and dispersion state of the nanoparticles were studied. Four different kinds of Cu–Sn base coatings, Cu–Sn, Cu–Sn–PTFE, Cu–Sn–TiO_2 and Cu–Sn–PTFE–TiO_2, were obtained and their properties were evaluated and compared, especially in terms of tribological and anti-corrosion properties.

2. Experimental Section

2.1. Electroplating Processes and Methods

Stainless steel 9Cr18 (Huatai, Yangzhou, China) was chosen as a cathode substrate with dimensions of 28 mm × 25 mm × 1 mm. Prior to electroplating, the substrate was ground by using CW (CW, type of abrasive paper made in Yuli, Xianning, China) series sand paper of 1500 grade, then degreased ultrasonically in alkaline and acid solution alternately.

During electroplating, a unidirectional pulse current from Shenzhen Shicheng Electronic Technology Co., Ltd. (Shenzhen, China) was used. The main parameters involve average current (I_a = 25 mA/cm^2), pulse frequency (f = 2000 Hz) and duty cycle (θ = 60%). The corresponding current-on times (t_{on}), current-off times (t_{off}) and peak current density (I_p) are respectively 3 ms, 2 ms and 41.7 mA/cm^2. Many experiments were carried out to obtain appropriate particle concentration of PTFE emulsion and TiO_2 sol [20]. On the basis of previous experiments, the concentration of PTFE and TiO_2 were determined at 15 and 1 g/L, respectively. The plating solution compositions and technological parameters are shown in Table 1.

Table 1. Plating solution compositions and technological parameters.

Compositions/Parameters	Quantity
$K_4P_2O_7 \cdot 3H_2O$	260–270 g/L
$Cu_2P_2O_7 \cdot 4H_2O$	20 g/L
$KNaC_4H_4O_6 \cdot 4H_2O$	30–35 g/L
$Na_2SnO_3 \cdot 3H_2O$	40 g/L
KNO_3	40 g/L
$Na_3C_6H_5O_7 \cdot 2H_2O$	20 g/L
PTFE	15 g/L
TiO_2 sol	1 g/L
time	1 h
pH	9–10
speed	100 r/min
temperature	35–40 °C

In the plating solution, the main salt ions are Cu^{2+} and SnO_3^{2-}. Potassium pyrophosphate ($K_4P_2O_7 \cdot 3H_2O$) is the main complexing agent, which provides $P_2O_7^{4+}$ to cause a complexation reaction with Cu^{2+} and Sn^{2+}. This transforms the discharge ion of tin from SnO_3^{2-} to $[SnP_2O_7^{4+}]^{4-}$ and promotes the co-deposition of copper and tin. Potassium sodium tartrate ($KNaC_4H_4O_6 \cdot 4H_2O$) is an auxiliary coordination agent, which can prevent the precipitation of copper hydroxide and the hydrolysis of stannate. Simultaneously, with relatively positive discharge potential, potassium nitrate (KNO_3) as a depolarizing agent can effectively reduce the polarization of the cathode and significantly promote tin deposition. Similarly, sodium citrate ($Na_3C_6H_5O_7 \cdot 2H_2O$) is an additive, which can also reduce the cathode polarization and indent the deposition potential of Cu^{2+} and Sn^{4+}.

The concentrations of PTFE and TiO_2 during deposition in each plating solution are shown in Table 2. In the bath solution, the Cu^{2+}, Sn^{4+}, PTFE and TiO_2 particles all have a positive charge. Under the influence of an electric field force, the particles move towards the cathode phase and are adsorbed on the cathode material before, finally, the Cu–Sn–PTFE–TiO_2 coating forms. The formation process is shown in Figure 1.

Table 2. The type and quantity of composite particles in each plating solution.

Content	Cu–Sn	Cu–Sn–PTFE	Cu–Sn–TiO_2	Cu–Sn–PTFE–TiO_2
PTFE (g/L)	0	15	0	15
TiO_2 (g/L)	0	0	1	1

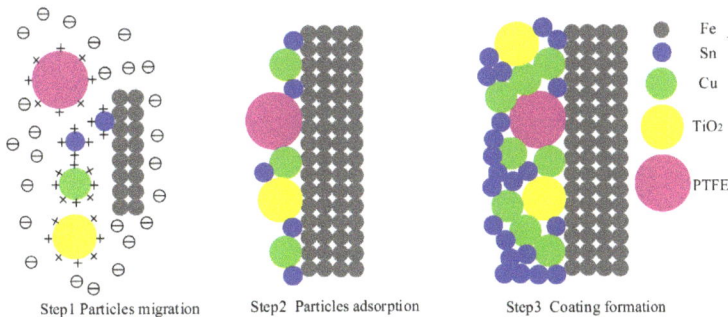

Figure 1. The schematic diagram of the Cu–Sn–PTFE–TiO_2 coating formation process (note that the relative sizes of ions/atoms/particles are not to scale).

2.2. Characterization of Plating Solution and Cu–Sn–PTFE–TiO$_2$ Composite Coatings

A 380ZLS model laser particle size/potential meter (Particle Sizing Systems, Port Richey, FL, USA) was used to measure the size of the TiO$_2$ sol and PTFE particles. In addition, the plating solutions of Cu–Sn–PTFE, Cu–Sn–TiO$_2$ and Cu–Sn–PTFE–TiO$_2$ were observed by transmission electron microscopy (TEM, FEI, Hillsboro, OR, USA). The samples for the TEM observation were made in the following way: take a drop of the solution using a plastic dropper and spread it on a micro-grid of carbon film on a copper mesh, then heat it to bake off the alcohol and water using an electric baking lamp. The surface morphologies and compositions of coatings were investigated with scanning electron microscopy (SEM, FEI) and X-ray Fluorescence (XRF, PANalytical, Almelo, the Netherlands). Hardness of the coatings, an average of five indentations, was measured by Vickers Hardness Tester HVS-1000 (Jujing, Shanghai, China) with a load of 2 N for 15 s on the surface.

Tribological properties of coatings were characterized by a friction–abrasion testing machine in room temperature. The friction counterpart was a GCr15 steel ball of 5 mm in diameter. A load of 100 g and a stage rotated speed of 200 r/min were used with a wear time of 10 min for each sample. Finally, the corrosion resistance of coatings was analyzed by Tafel polarization tests, which were conducted in a three-electrode system. The counter electrode was a platinum slice, and the reference electrode was a standard calomel electrode (SCE). The tests were performed in 3.5% NaCl electrolyte with an electrochemical workstation (model CHI660B, Chenhua, Shanghai, China) at a scan rate of 2 mV/s. The electrochemical behavior of the plating solution was also tested with this workstation. The volt–ampere characteristic curves were obtained by linear sweep voltammetry (LVS).

3. Results and Discussion

3.1. Analysis of Plating Solution

3.1.1. Electrodepositional Behavior of Basic Solution

Figure 2 shows the volt–ampere characteristic curves (LSV) of the solutions, including K$_4$P$_2$O$_7$. The LSV of the solution with just K$_4$P$_2$O$_7$·3H$_2$O (260 g/L) is shown in Figure 2a, which shows the current rising gradually with the potential negative shift. Although hydrogen evolution starts to occur when the potential goes negative to −1.6 V, there is no other reduction peak during the scanning range (−0.2~−1.6 V) and the solution exhibits good stability. The LSV of the solutions with Cu$_2$P$_2$O$_7$·4H$_2$O (20 g/L) are shown in Figure 2b. It can be seen that the cathode current starts to rise at −0.4 V due to the reduction of Cu^{2+} ions to Cu metal (curve a in Figure 2b). With the addition of K$_4$P$_2$O$_7$·3H$_2$O (260 g/L), cathode potential of Cu shifts to approximately −1.0 (curve b in Figure 2b).

The LSV of Sn deposition is shown in Figure 2c. With Na$_2$SnO$_3$·3H$_2$O (40 g/L) and K$_4$P$_2$O$_7$·3H$_2$O (260 g/L) in the solution, the deposition of Sn^{2+} ions to Sn metal starts to occur at −1.25 V. With the addition of K$_4$P$_2$O$_7$·3H$_2$O (260 g/L), two changes occur: a more positive potential −1.0 V and an increased peak. So, with K$_4$P$_2$O$_7$·3H$_2$O in the solution, the deposition of Sn seems to become easy. Reduction peak is observed at −0.9 V from the LSV of the solution containing K$_4$P$_2$O$_7$·3H$_2$O (260 g/L), Cu$_2$P$_2$O$_7$·4H$_2$O (20 g/L) and Na$_2$SnO$_3$·3H$_2$O (40 g/L), shown in Figure 2d. In contrast with Figure 2b,c, the gap between the reduction potentials of Cu and Sn is found to be diminished with the presence of K$_4$P$_2$O$_7$·3H$_2$O. Therefore, the inclusion of K$_4$P$_2$O$_7$·3H$_2$O facilitates the co-deposition by lowering the difference in reduction potentials of the two individual metals (see Figure 2d).

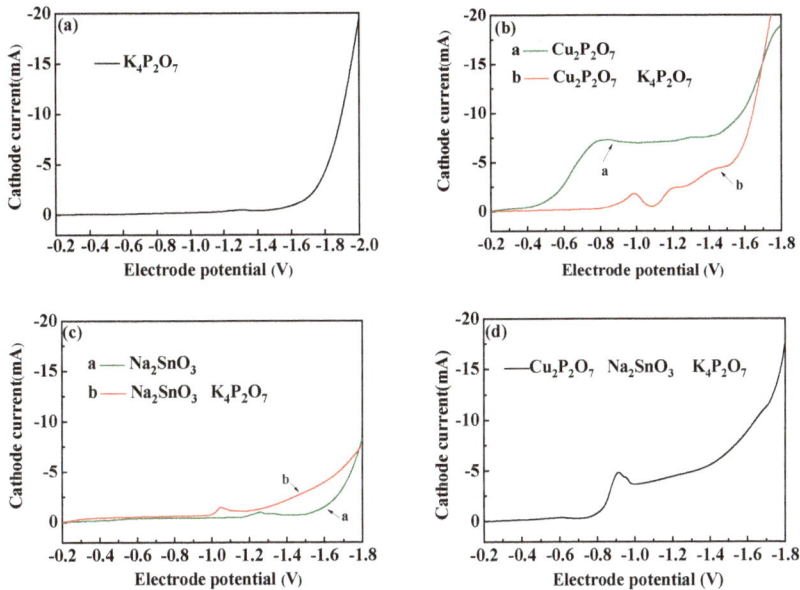

Figure 2. The volt–ampere characteristic curves of solutions: (**a**) $K_4P_2O_7$ solution; (**b**) electrolytic deposition of Cu^{2+}; (**c**) electrolytic deposition of Sn^{4+}; (**d**) electrolytic deposition of Cu^{2+} and Sn^{4+}.

3.1.2. Analysis of TiO$_2$ and PTFE particles

Figure 3 shows the distribution ranges of particle sizes and average diameters of the PTFE and TiO$_2$. While both particle size distributions show log-normal distributions (as expected), the left cut-off in Figure 3a (PTFE particles) is caused by the detection limit of the particle analyzer. In other words, particles < 25 mm could not be detected. Thus, it can be concluded that the average particle diameters of PTFE are less than 283 nm. Figure 3b shows that the average particle diameters of TiO$_2$ is 158 nm.

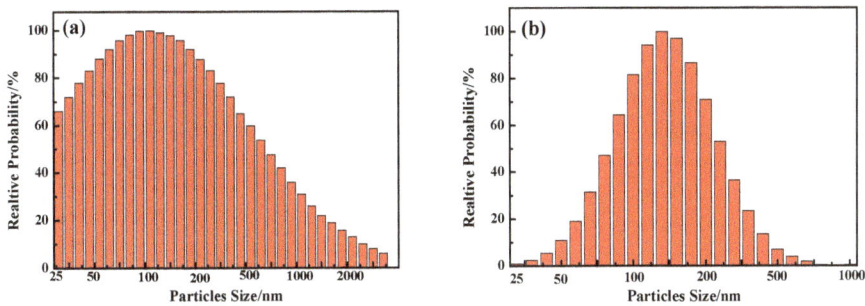

Figure 3. The size distribution of PTFE emulsion and TiO$_2$ sol: (**a**) PTFE emulsiom; (**b**) TiO$_2$ sol (note that the detection limit of the particle analyzer is 25 mm).

The obtained PTFE emulsion and TiO$_2$ sol particles were added into the base plating solution for further testing. With the addition of 15 g/L PTFE and 1 g/L TiO$_2$, respectively, the particles were uniformly dispersed in the plating solution and no agglomeration occurred, as observed in Figure 4a,b. Additionally, the particle size was not significantly changed (compared with Figure 3). When PTFE and TiO$_2$ were added to the solution together, the solution remained steady and the dispersity of the nanoparticles was still well (Figure 4c).

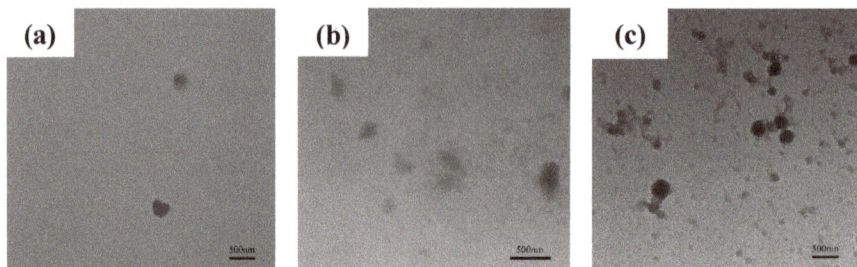

Figure 4. The dispersion state morphologies of PTFE emulsion and TiO$_2$ sol in electroplating solution: (**a**) PTFE emulsion; (**b**) TiO$_2$ sol; (**c**) PTFE–TiO$_2$.

3.2. Microstructure and Composition of the Coatings

The SEM micrographs of coatings, including Cu–Sn, Cu–Sn–PTFE, Cu–Sn–TiO$_2$ and Cu–Sn–PTFE-TiO$_2$ composite coatings are shown in Figure 5. According to SEM results, the surface of each coating is uniform and there are no such defects as pinhole, cracks, drain plating, etc. With the addition of TiO$_2$, the coatings of Cu–Sn–TiO$_2$ and Cu–Sn–PTFE–TiO$_2$ are smoother than Cu–Sn and Cu–Sn–PTFE. Indeed, the results also show TiO$_2$ sol has dramatically effect on improving the coating structure than that of PTFE emulsion. However, with PTFE and TiO$_2$ coexisting in the coating, Cu–Sn–PTFE–TiO$_2$ composite coating possess the smoothest surface and finest structure.

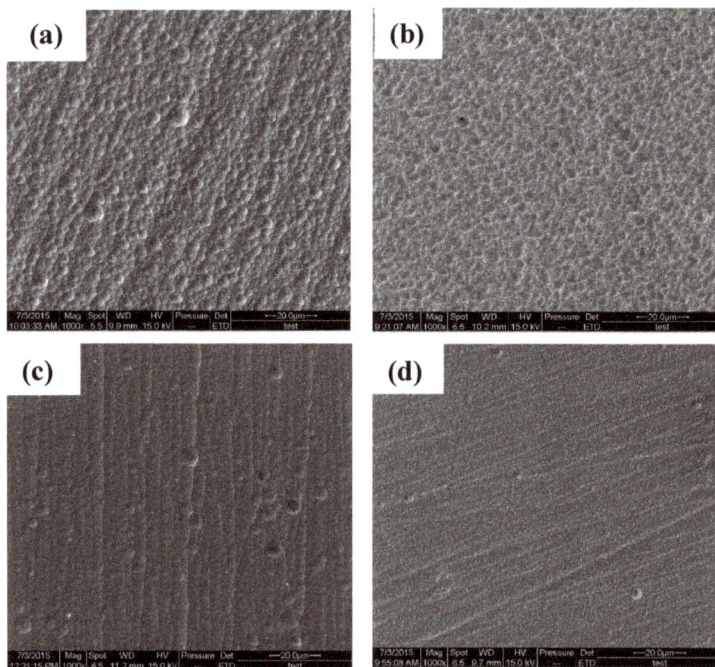

Figure 5. The SEM morphologies of Cu–Sn base coatings: (**a**) Cu–Sn; (**b**) Cu–Sn–PTFE; (**c**) Cu–Sn–TiO$_2$; (**d**) Cu–Sn–PTFE–TiO$_2$.

Table 3 shows the composition of the deposits produced with different particles in the bath. According to the XRF analysis, the alloy composition is dependent on the composition of the bath. Although both of the contents of Cu and Sn decrease with the addition of PTFE and TiO_2, the proportions of Cu/Sn in the coatings stay basically unchanged. Composition analysis also confirms that PTFE and TiO_2 coexist in the coating of Cu–Sn–PTFE–TiO_2.

Table 3. The compositions of the coatings.

Conversion Coatings	Cu (wt %)	Sn (wt %)	PTFF (wt %)	TiO$_2$ (wt %)
Cu–Sn	92.42	7.58	—	—
Cu–Sn–PTFE	87.21	6.68	6.11	—
Cu–Sn–TiO$_2$	90.71	6.76	—	1.54
Cu–Sn–PTFE–TiO$_2$	86.84	6.46	5.50	1.20

3.3. Corrosion Properties

The potentiodynamic polarization curves of different kinds of Cu–Sn base composite coatings are shown in Figure 6. Table 4 presents the corresponding corrosion current density and corrosion potential. The corrosion resistances of Cu–Sn–PTFE, Cu–Sn–TiO_2 and Cu–Sn–PTFE–TiO_2 coatings are much better than that of the Cu–Sn alloy coating in 3.5% NaCl solution. Cu–Sn–PTFE–TiO_2 composite coating also has the highest corrosion potential and minimum corrosion current density, which are -0.256 V and 1.443 A/cm^2, respectively.

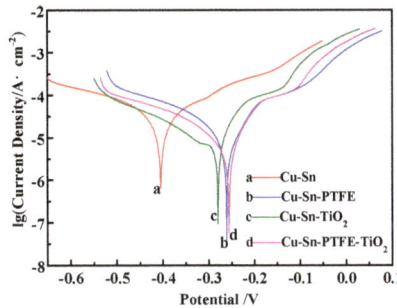

Figure 6. Potentiodynamic polarization curves of different coatings.

Table 4. Potentiodynamic polarization parameters of coatings.

Coatings	Corrosion Potential/V	Corrosion Current Density/A·cm^{-2}
Cu–Sn	-0.405	4.067
Cu–Sn–PTFE	-0.261	1.856
Cu–Sn–TiO$_2$	-0.280	2.434
Cu–Sn–PTFE–TiO$_2$	-0.256	1.443

Figure 7 presents the corroded surface topographies of four kinds of coatings. The main corrosion form of all of them was point corrosion in 3.5% NaCl electrolyte. However, the depth of the corrosion pits was significantly different in those coatings. It is obvious that Cu–Sn–PTFE–TiO_2 has the shallowest corrosion pits and the smoothest surface.

In any case, with the same corrosion medium and corrosion time, the anti-corrosion performances of Cu–Sn–PTFE, Cu–Sn–TiO_2, Cu–Sn–PTFE–TiO_2 coatings were distinctly better than that of Cu–Sn alloy coatings. Presumably the main reason is that the pulses co-deposited of PTFE and TiO_2 nanoparticles produce fine grains, which strengthen matrix structure, compact the microstructure and improve the corrosion resistance. In addition, PTFE and TiO_2 nanoparticles can fill in the intergranular pores of matrix grain, and the reduction of the pore size on the material surface can prevent the corrosion ions getting through micropores of the composite material, so the anti-corrosion performance is effectively improved (Table 4).

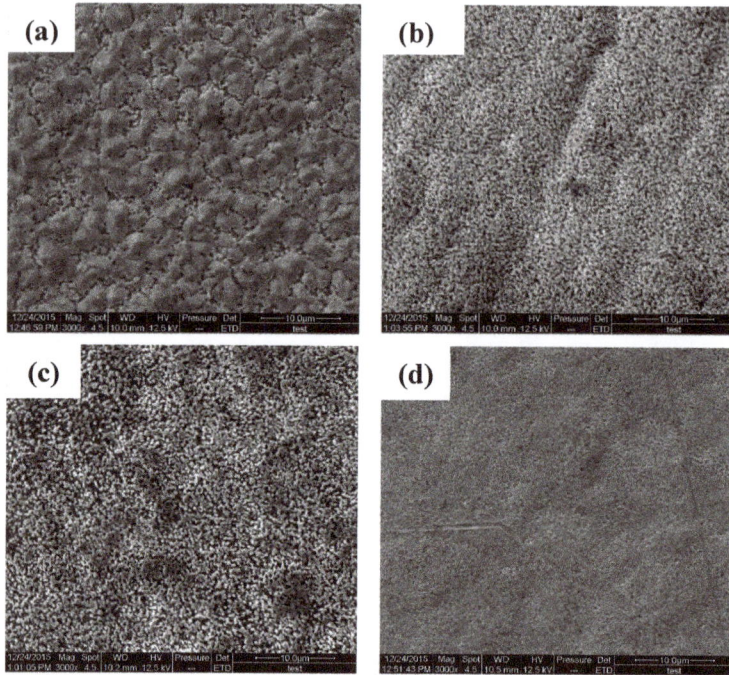

Figure 7. The SEM morphologies of corroded coatings: (**a**) Cu–Sn; (**b**) Cu–Sn–PTFE; (**c**) Cu–Sn–TiO$_2$; (**d**) Cu–Sn–PTFE–TiO$_2$.

3.4. Mechanical Properties

3.4.1. Hardness of Coatings

The hardness values of different kinds of Cu–Sn base composite coatings are listed in Table 5. The results suggest that the hardness of Cu–Sn–TiO$_2$, Cu–Sn–PTFE–TiO$_2$, Cu–Sn and Cu–Sn–PTFE coatings rank from high to low, indicating that TiO$_2$ nanoparticles are able to increase the hardness of the coating, while PTFE particles decrease. In the electrodeposition process, the hard TiO$_2$ nanoparticles were embedded in the Cu–Sn alloy matrix, which lead to higher hardness than Cu–Sn coating. That could presumably be interpreted as fine-grain strengthening and dispersion strengthening effect. However, due to the softness of PTFE particles, hardness of the Cu–Sn–PTFE coating is lower than that of Cu–Sn coating. With the synergistic effect of TiO$_2$ and PTFE, the Cu–Sn–PTFE–TiO$_2$ composite coating can achieve higher hardness than Cu–Sn coating.

Table 5. Hardness of the coatings.

Coating	Hardness/HV
Cu–Sn	413 ± 4
Cu–Sn–PTFE	375 ± 6
Cu–Sn–TiO$_2$	485 ± 6
Cu–Sn–PTFE–TiO$_2$	465 ± 5

3.4.2. Tribological Properties

The friction coefficient curves of different kinds of those coatings are presented in Figure 8. It indicates that under dry friction conditions, the Cu–Sn alloy coating exhibits the maximum friction coefficient along with poor wear resistance. According to Figure 8, PTFE decreases the friction coefficient to 0.08, but cannot improve the wear life. On the other hand, with the addition of nano-TiO_2, wear-resisting time of the Cu–Sn–TiO_2 composite coating increases, while the friction coefficient is about 0.14 between those of Cu–Sn and Cu–Sn–PTFE. However, the Cu–Sn–PTFE–TiO_2 composite coating has much longer wear-resisting time, as same as Cu–Sn–TiO_2. In addition, its friction coefficient of 0.1 is the lowest among the four kinds of coatings. Therefore, the Cu–Sn–PTFE–TiO_2 coating achieves the best wear-resisting and anti-friction performance with PTFE and TiO_2 coexisting.

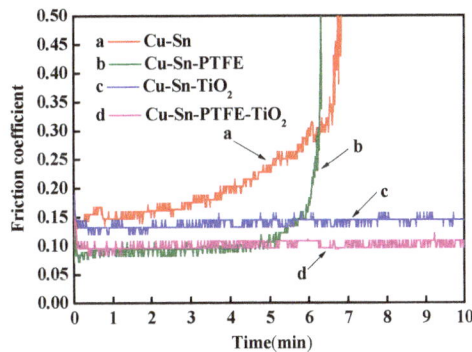

Figure 8. The friction coefficient curve of each Cu–Sn base plating coating.

The SEM images of the worn surfaces of those composite coatings are shown in Figure 9. Severe plastic deformation happens on the worn surface of the Cu–Sn alloy and Cu–Sn–PTFE coating (Figure 9a,b), which is usually related to adhesive wear mechanism, while narrow scratches implying abrasive wear mechanism are observed in the worn surface of the Cu–Sn–TiO_2 and Cu–Sn–PTFE–TiO_2 coatings (Figure 9c,d). Simultaneously, with the addition of TiO_2 (Figure 9c,d), wear tracks become smoother and their widths also become narrower than those of coatings without TiO_2 (Figure 9a,b). However, just some slight furrowing or scratches appear on the surface of the Cu–Sn–PTFE–TiO_2 composite coating, which indicate the best wear resistance (Figure 9d).

The wear mechanisms of electroplated coatings against steel can be explained in Figure 10. Steel balls with a higher hardness of HV800 were used as counterparts. When a Cu–Sn deposit with lower hardness was chosen as the tested sample, shown in Figure 10a, the small steel ball would be easily embedded in the coating matrix in the process of relative sliding, which made the contact area and friction force increase and result in serious wear damage. Once the tested sample was changed to a Cu–Sn–PTFE composite coating, as shown in Figure 10b, the friction shear force was significantly reduced with the formation of PTFE self-lubricating film, however, wear resistance of the coating was not improved due to the low hardness of the coating.

In contrast, a relatively high hardness brought by the reinforcement of TiO_2 nanoparticles increases the resistance of plastic distortion in the process of relative motion, plus no self-lubricating film on the surface, which leads to a higher friction coefficient (Figure 10c). However, with PTFE and TiO_2 coexisting in the Cu–Sn–PTFE–TiO_2 composite coating, as illustrated in Figure 10d, lower shear force and harder matrix correspondingly lead to a lower coefficient and higher wear resistance. That is to say, the synergistic effect of PTFE and TiO_2 makes the Cu–Sn–PTF–TiO_2 composite coating present a good wear-resisting and anti-friction performance.

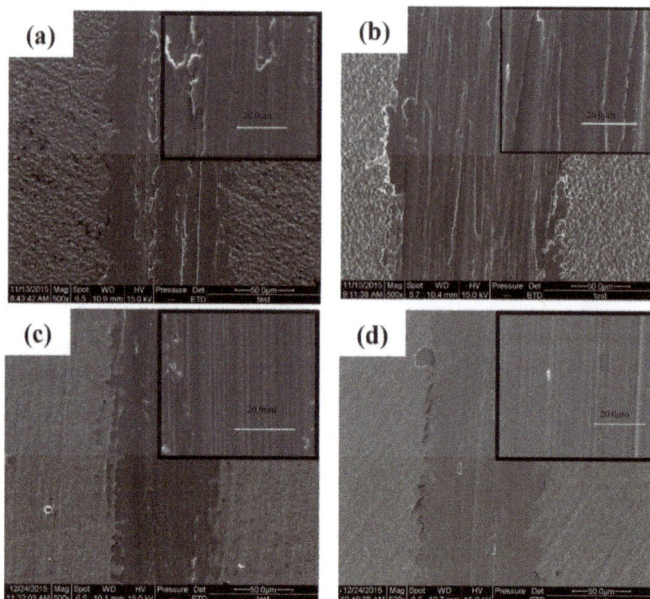

Figure 9. The SEM morphologies of wear traces on the surface of coatings: (**a**) Cu–Sn; (**b**) Cu–Sn–PTFE; (**c**) Cu–Sn–TiO$_2$; (**d**) Cu–Sn–PTFE–TiO$_2$.

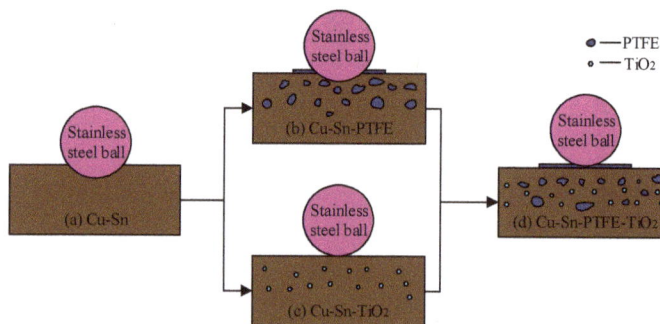

Figure 10. The schematic diagram of the friction mechanism of each Cu–Sn base coatings: (**a**) Cu–Sn; (**b**) Cu–Sn–PTFE; (**c**) Cu–Sn–TiO$_2$; (**d**) Cu–Sn–PTFE–TiO$_2$.

4. Conclusions

In the paper, PTFE emulsion and TiO$_2$ sol with average particle diameters <283 nm and ~158 nm, respectively, were successfully prepared and dispersed. TEM analysis indicates that no agglomeration occurs in the plating solution with appropriate particle concentrations. The analysis of the electrochemical behavior suggests that Cu and Sn can be co-deposited relatively easily with the addition of K$_4$P$_2$O$_7$. Hence, Cu–Sn composite coatings containing PTFE and TiO$_2$ are obtained from a pyrophosphate plating solution, which has superior antiwear and friction reduction properties.

The investigations also indicate that PTFE and TiO$_2$ nanoparticles are able to promote the coating structure and smoothness. Simultaneously, with the addition of nanoparticles, the corrosion resistance of Cu–Sn–PTFE, Cu–Sn–TiO$_2$ and Cu–Sn–PTFE–TiO$_2$ composite coatings is better than that of Cu–Sn alloy coating. Cu–Sn–PTFE–TiO$_2$ composite coating also has the highest corrosion potential and minimum corrosion current density, which are -0.256 V and 1.443 A/cm^2, respectively.

In addition, the hardness of the Cu–Sn–TiO$_2$ coating is 485 HV, while that of the Cu–Sn–PTFE–TiO$_2$ coating is 465 HV due to the softness of PTFE. However, with the synergistic effect of PTFE and TiO$_2$, a Cu–Sn–PTFE–TiO$_2$ composite coating exhibits superior wear-resisting and anti-friction performance with the friction coefficient of 0.10.

Author Contributions: Conceptualization, L.Y.; Methodology, L.Y. and K.W.; Validation, K.W., Z.F. and C.W.; Formal Analysis, Z.F., K.W.; Investigation, Z.F., K.W.; Data Curation, L.Y., Z.F. and K.W.; Writing–Original Draft Preparation, Z.F. and K.W.; Writing–Review & Editing, L.Y., C.W., T.Z., Y.X. and G.W.

Funding: This research was funded by the National Nature Science Foundation of China (No.51305090) and the Fundamental Research Funds for the Central Universities (HEUCFP201814).

Conflicts of Interest: The authors declare no conflict of interest. The funders had no role in the design of the study; in the collection, analyses, or interpretation of data; in the writing of the manuscript, and in the decision to publish the results.

References

1. Pewnim, N.; Roy, S. Electrodeposition of tin-rich Cu–Sn alloys from a methanesulfonic acid electrolyte. *Electrochim. Acta* **2013**, *90*, 498–506. [CrossRef]
2. Zanella, C.; Xing, S.; Deflorian, F. Effect of electrodeposition parameters on chemical and morphological characteristics of Cu–Sn coatings from a methanesulfonic acid electrolyte. *Surf. Coat. Technol.* **2013**, *236*, 394–399. [CrossRef]
3. Subramanian, B.; Mohan, S.; Jayakrishnan, S. Structural, microstural and corrosion properties of brush plated Copper–Tin alloy coatings. *Surf. Coat. Technol.* **2006**, *201*, 1145–1151. [CrossRef]
4. Bengoa, L.N.; Pary, P.; Conconi, M.S.; Egli, W.A. Electrodeposition of Cu–Sn alloys from a methanesulfonic acid electrolyte containing benzyl alcohol. *Electrochim. Acta* **2017**, *256*, 211–219. [CrossRef]
5. Cui, G.; Bi, Q.; Zhu, S.; Yang, J.; Liu, W. Tribological properties of bronze–graphite composites under sea water condition. *Tribol. Int.* **2012**, *53*, 76–86. [CrossRef]
6. Mallikarjuna, H.M.; Kashyap, K.T.; Koppad, P.G.; Ramesh, C.S.; Keshavamurthy, R. Microstructure and dry sliding wear behavior of Cu–Sn alloy reinforced with multiwalled carbon nanotubes. *Trans. Nonferr. Met. Soc. China* **2016**, *26*, 1755–1764. [CrossRef]
7. Balaji, R.; Pushpavanam, M.; Kumar, K.Y.; Subramanian, K. Electrodeposition of bronze–PTFE composite coatings and study on their tribological characteristics. *Surf. Coat. Technol.* **2006**, *201*, 3205–3211. [CrossRef]
8. Du, C.J.; Li, W.W.; Guo, D.P.; Liu, X.G.; Wang, Y.L.; Zhang, L.; Nan, Y. Electrodeposition of Cu–Sn–Zn–PTFE composite coating and its self-lubricating property. *Electroplat. Pollut. Control.* **2017**, *5*, 12–14. (In Chinese)
9. Banerjee, A.; Kumar, R.; Dutta, M.; Bysakh, S.; Bhowmick, A.K.; Laha, T. Microstructural evolution in Cu–Sn coatings deposited on steel substrate and its effect on interfacial adhesion. *Surf. Coat. Technol.* **2015**, *262*, 200–209. [CrossRef]
10. Low, C.T.J.; Wills, R.G.A.; Walsh, F.C. Electrodeposition of composite coatings containing nanoparticles in a metal deposit. *Surf. Coat. Technol.* **2006**, *201*, 371–383. [CrossRef]
11. Wu, Y.; Liu, L.; Shen, B.; Hu, W. Investigation in electroless Ni–P–C (graphite)–SiC composite coating. *Surf. Coat. Technol.* **2006**, *201*, 441–445. [CrossRef]
12. Su, F.; Zhang, S. Tribological properties of polyimide coatings filled with PTFE and surface-modified nano-Si$_3$N$_4$. *J. Appl. Polym. Sci.* **2014**, *131*, 383–390. [CrossRef]
13. Sieh, R.; Le, H. Non-cyanide electrodeposited Ag–PTFE composite coating using direct or pulsed current deposition. *Coatings* **2016**, *6*, 31–44. [CrossRef]
14. Xu, X.; Yao, W.; Xu, J.; Zhang, W.; Yang, L.; Deng, S. The influences of the plating's hardness, abrasion and antifriction property caused by the adding of nano-Al$_2$O$_3$ and PTFE. *Adv. Mater. Res.* **2012**, *517*, 1683–1686. [CrossRef]
15. Suiyuan, C.; Ying, S.; Hong, F.; Jing, L.; Liu, C.S.; Sun, K. Synthesis of Ni–P–PTFE-nano-Al$_2$O$_3$ composite plating coating on 45 steel by electroless plating. *J. Compos. Mater.* **2012**, *46*, 1405–1416. [CrossRef]
16. Chen, W.; He, Y.; Gao, W. Synthesis of nanostructured Ni–TiO$_2$ composite coatings by sol-enhanced electroplating. *Electrochem. Soc.* **2010**, *157*, E122–E128. [CrossRef]

17. Chen, W.; Gao, W. Thermal stability and tensile properties of sol-enhanced nanostructured Ni–TiO$_2$ composites. *Compos. Part A Appl. Sci. Manuf.* **2011**, *42*, 1627–1634. [CrossRef]
18. Wang, Y.; Wang, S.; Shu, X.; Gao, W.; Lu, W.; Yan, B. Preparation and property of sol-enhanced Ni–B–TiO$_2$ nano-composite coatings. *J. Alloy. Compd.* **2014**, *617*, 472–478. [CrossRef]
19. Wang, Y.; Ju, Y.; Wei, S.; Gao, W.; Lu, W.; Yan, B. Au–Ni–TiO$_2$ nano-composite coatings prepared by sol-enhanced method. *J. Electrochem. Soc.* **2014**, *161*, 775–781. [CrossRef]
20. Ying, L.; Li, Z.; Wu, K.; Lv, X.; Wang, G. Effect of TiO$_2$ sol on the microstructure and tribological properties of Cu–Sn coating. *Rare Met. Mater. Eng.* **2017**, *46*, 2801–2806. [CrossRef]

coatings

MDPI

Article

Assessment the Sliding Wear Behavior of Laser Microtexturing Ti6Al4V under Wet Conditions

Juan Manuel Vazquez Martinez [1,*], Irene Del Sol Illana [1], Patricia Iglesias Victoria [2] and Jorge Salguero [1]

[1] Department of Mechanical Engineering & Industrial Design, Faculty of Engineering, University of Cadiz, Av. Universidad de Cadiz 10, E-11519 Puerto Real-Cadiz, Spain; irene.delsol@uca.es (I.D.S.I.); jorge.salguero@uca.es (J.S.)

[2] Department of Mechanical Engineering, Rochester Institute of Technology, 72 Lomb Memorial Drive, Rochester, NY 14623, USA; pxieme@rit.edu

* Correspondence: juanmanuel.vazquez@uca.es; Tel.: +34-956-483-513

Received: 19 December 2018; Accepted: 22 January 2019; Published: 24 January 2019

Abstract: Laser micro-texturing processes, compared to untreated surfaces, can improve the friction, wear and wettability behavior of sliding parts. This improvement is related to the micro-geometry and the dimensions of the texture which is also dependent on the processing parameters. This research studied the effect of laser textured surfaces on the tribological behavior of titanium alloy Ti6Al4V. The influence of processing parameters was analyzed by changing the scanning speed of the beam and the energy density of pulse. First, the characterization of dimensional and geometrical features of the texturized tracks was carried out. Later, their influence on the wetting behavior was also evaluated through contact angle measurements using water as a contact fluid. Then, the tribological performance of these surfaces was analyzed using a ball-on-flat reciprocating tribometer under wet and dry conditions. Finally, wear mechanisms were identified employing electronic and optical microscopy techniques capable to evaluate the wear tracks on Ti surfaces and WC–Co spheres. These analyses had determined a strong dependence between the wear behavior and the laser patterning parameters. Wear friction effects were reduced by up to a 70% replacing conventional untreated surfaces of Ti6Al4V alloy with laser textured surfaces.

Keywords: surface modification; laser texturing; wetting behavior; Ti6Al4V; surface characterization; tribology; wear behavior

---

1. Introduction

Surface modification of metallic surfaces has shown an important growth of engineering applications in recent years. With relevant developments in the fields of aerospace, energy and biomedical industries. The surface modification based on texturing processes is mainly employed for enhancing physic-chemical properties of the substrate material [1–8]. This kind of techniques represents an essential tool to improve the functional performance of specific alloys in working environments, through the variation of specific features as wetting or tribological behavior [9–15]. By using laser surface processing procedures regular or random texture patterns can be achieved [13,16–21]. Laser texturing techniques can be used to induce a large variety of topographies in terms of the distribution and geometry of the irradiated pattern being the most common linear, circular and pulses array distributions [13,16,19,21]. Due to this fact, it is known that the geometrical and dimensional characteristics of the generated structures have a direct influence with the contact angle between the liquid and solid phases, resulting in higher or lower water retention capability [13,19,22–26].

Texturing geometries may act as wear debris traps and water reservoirs, thereby improving the performance under tribological applications of surfaces. Under these conditions, these textures mainly help remove wear debris from the sliding track, also reducing the involved abrasive effects of the detached particles. Moreover, laser processing shows some relevant advantages regarding chemical and mechanical processes. On one hand, the laser system can be adapted to a wide range of parameters for the texturing process, become a more flexible tool than machining operations for surface modification operations. These kinds of treatments, based on laser irradiation under air atmosphere, are usually performed without lubricants or coolant fluids, being environmentally friendly regarding chemical and conventional machining [2,27–29].

Ultra short laser pulses are typically used for metal surface texturing, resulting in melting and vaporizing the material. This technique is mainly employed to remove selective material and obtaining well defined textured patterns. An additional benefit of the ultra-short laser pulses (i.e., picosecond or even femtosecond) is the prevention of material pile-up around the edges of the textured features (pockets). However, this type of laser systems has important disadvantages, such as low resistance to industrial environments and high costs, which makes it difficult to integrate into real life applications [21,30–32]. However, they have been recently integrated into the manufacturing process for specific applications [33,34]. The use of conventional laser marking systems for the textures development, usually rejected for this purpose due to their requirement of use in pulse mode, imply a high understanding of the influence of laser processing parameters on the resulting texture features. For linear textures, energy density of pulse (E_d) and scanning speed of the beam (V_s) are shown as two of the most relevant parameters involved in the size and geometrical properties of the laser grooves [35–38].

Titanium alloys are commonly used, among others, for the manufacturing of aerospace parts or biomedical components mainly due to their good ratio between weight and mechanical properties and excellent biocompatibility. The use of laser texturing treatments is focused to overcome functioning limitations, such as poor wear behavior. The unstable frictional response of titanium alloys, coupled with severe wear behavior under certain rubbing conditions, make the use of this type of alloys difficult to apply in real scenario tribological applications [39–41]. Under this consideration, the ability to obtain better liquid retention is showed as one of the most effective ways to improve the friction and sliding behavior. Through the development of textured topographies with specific geometry and size (the retention of water over the surface can be obtained) is possible to maintain control of the liquid film thickness. These effects may be mainly obtained by variations on the roughness of the modified surface layer, also controlled by combinations of laser parameters of the process. Moreover, the manufacture of laser microgrooves can be used as micro-channels for controlling direction and flow rate, additionally improving the corrosion resistance and lubricant conditions [13,42–45]. Additionally, texturing geometries may act as wear debris traps and water reservoirs, thereby improving the performance under tribological applications of surfaces [46].

Based on this background and taking as starting point preliminary studies focused on the water retention of surfaces by microtexturing treatments [19], the aim of this research is to investigate the influence of laser processing parameters on the improvement of tribological wear behavior of Ti6Al4V surfaces under wet conditions. A laser surface treatment using a conventional laser marking system have been presented to develop specific surface patterns of linear tracks on titanium samples. Through variations of E_d and V_s a wide range of topographies with different features have been generated, giving place to variations in the contact angle of water droplets. These water retention changes, mainly achieved through surface finishing, can also impact the frictional response and wear behavior.

2. Materials and Methods

2.1. Laser Texturing Process

Initially the base surface roughness was set to $R_a < 0.05$ μm and $R_z < 0.15$ μm by mechanical polishing and cleaned by ether-petroleum 50% dissolution. Textured specimens of 10×10 mm^2 side and 5 mm thickness were performed on Ti6Al4V titanium plates using a using a Ytterbium fiber infrared laser (1070 ± 5 nm wavelength), model Rofin EasyMark F20 system (ROFIN-SINAR Technologies Inc., Plymouth, MI, USA). Spot diameter and pulse width were fixed to 60 μm and 100 ns respectively. The treatments were performed under an open-air atmosphere. Textured surfaces were developed through bidirectional parallel lines. There was no overlapping and the distance between the laser tracks was 0.1 mm.

In order to evaluate the influence of laser processing parameters on the variation of the properties and features of the textured surfaces, different treatments conditions were studied. Three different energy density of pulse (E_d) were chosen and combined with eight scanning speed of the beam (V_s), values are shown in Table 1. In total, 24 different textures were characterized for wettability behavior and surface roughness. 12 of these samples corresponding to 10, 40, 100 and 250 mm/s of V_s, plus the non-textured specimen, selected to perform the tribological tests.

Table 1. Laser processing parameters.

E_d (J cm^{-2})	V_s (mm s^{-1})
17.68–7.07–4.42	10–20–40–80–100–150–200–250

2.2. Textured Surfaces Characterization

The effects of texturing parameters are evaluated through the characterization of the treated surfaces.

Surface roughness measurements of all the textured surfaces were performed, choosing R_z and R_{Sm} [41] as control parameters. R_z parameter is defined as the maximum height between deeper and higher values of the asperities contained in the roughness profile during an evaluation length and R_{Sm} is described as the mean width of the profile elements [47]. These measurements were carried out employing a Mahr Perthometer Concept PGK120 (Mahr technology, Göttingen, Germany) surface profilometer. Scanning electron microscopy (SEM) techniques were also used to analyze the shape of the laser tracks. Additionally, the presence of oxidation on the customized surfaces was studied through energy dispersive spectroscopy (EDS) analysis.

The wetting properties of the modified surfaces play an important role in the improvement of the friction behavior under wet conditions [16,18]. For this reason, water retention ability of the different textures was evaluated measuring the contact angle between solid and liquid phases. A Ramé-Hart system (Ramé-hart Instrument Co., Succasunna, NJ, USA) and DropImage Advanced, as image processing software, were used for this purpose.

2.3. Pin-On-Flat Tribological Tests

Friction and sliding contact behaviors were evaluated by means of pin-on-flat reciprocating tests. Developed textures were tested against 1.5 mm diameter tungsten carbide balls. For each condition of the treated specimens, at least two repeated tests were performed applying a maximum Hertz contact pressure of 2.71 GPa, which corresponds to a medium Hertz contact pressure of 1.81 GPa. Total sliding distance and linear speed were fixed at 50 m and 0.012 m s^{-1} respectively. The reciprocating stroke length was initially set to 2 mm. All the reciprocating tests were conducted under dry and wet conditions. Water volume was set up on 15–20 μL for lubricating the laser textures in one single dose supplied at the beginning of the tests, and ensure a boundary condition regime.

2.4. Frictional Characterization of Test Specimens

Wear effects and dimensions on the sliding track, as well as the carbide balls used as pins, were evaluated by optical microscopy techniques using an Olympus SZX12 system (Olympus, Tokyo, Japan). Wear track and wear volume (V_f) caused by the sliding process on the modified surfaces were evaluated using Equation (1) [48] instead of using ASM standard [49]. This standard may not be recommended for textured samples due to its lack of surface uniformity and its difficulty obtaining a reference plane.

$$V_f = L_s \left[R^2 \times \sin^{-1} \left(\frac{W_t}{2 \times R_f} \right) - \left(\frac{W_t}{2} \right) (R_f - h_f) \right] + \pi \frac{(L - L_s)}{3W_t} \left[h_f^2 (3R_f - h_f) \right] \qquad (1)$$

where L_s is the stroke length, W_t is the average of 16 repeated measures of track width, L is the complete track length, R_f is the pin radio and h_f is the wear depth. In this case, h_f can be obtained by Equation (2).

$$h_f = R_f - \sqrt{R_f^2 - \frac{W_t^2}{4}} \qquad (2)$$

The wear tracks from all tribological tests were also analyzed employing a scanning electron microscopy (SEM) (Tescan Mira3, Brno–Kohoutovice, Czech Republic), and energy dispersive X-ray spectroscopy (EDX) (Tescan Mira3, Brno–Kohoutovice, Czech Republic) with the aim to evaluate the friction effects caused by the sliding motion. In which regards the worn pin area, the adhesion properties of the balls were characterized and measured through the image processing tools previously described. Adhesion shape was approximated to an ellipsoid to calculate it area.

3. Results and Discussion

3.1. Surface Finish of the Textures

The laser texturing process can create a modified layer with various topographies for specific applications. Through variations of laser processing parameters, the dimensions and geometry of the texturing tracks can be controlled. V_s was shown to be one of the main variables that govern laser surface modification in terms of microgeometrical texture generation, as can be observed in Figures 1 and 2.

The analysis of R_z measurements from the textured surfaces shows that a higher E_d may produce specific topographies with longer asperity dimensions. It does not seem to be a significant dependency or correlation between R_z and V_s. However, lower E_d ($E_d = 4.42$ J cm^{-2}; $E_d = 7.07$ J cm^{-2}) present an inflection point at $V_s = 150$ mm s^{-1}, where the decreasing trend changes to an increasing one. This singular point is caused as a result of the Ti cooling process; the Ti cooled down at the top of the surface obtruding the laser grooves and producing a smother surface. This obstruction is mainly due to a solidification stage in the surface section of vaporized material from the bottom of the textured groove [37].

Regarding R_{Sm} roughness parameter behavior, a direct influence of V_s was confirmed. As it can be observed in Figure 2, for laser texturing speeds V_s below 80–100 mm s^{-1}, a significant trend in the uniformity of the distance between peaks have been found. This effect is mainly caused by the lack of overlapping phenomena between laser textured tracks, allowing to maintain a constant distance between bidirectional parallel lines of the beam.

A significant decrease in the peak distance was detected for $E_d = 7.07$ J cm^{-2}, which is believed to be caused by the particular geometrical formation developed in the cooling process.

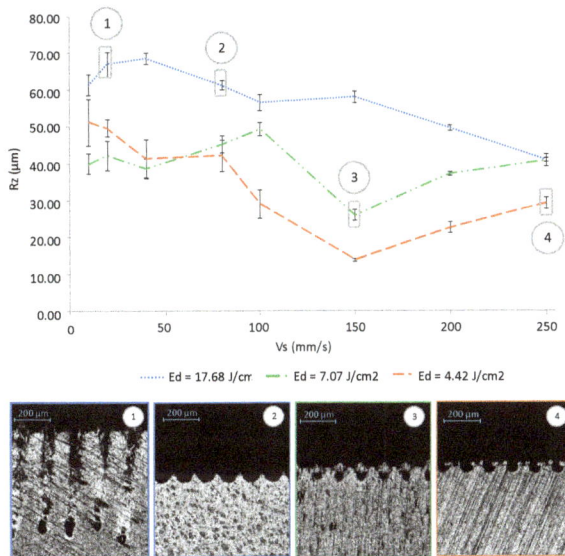

Figure 1. Roughness R_z behavior as a function of V_s.

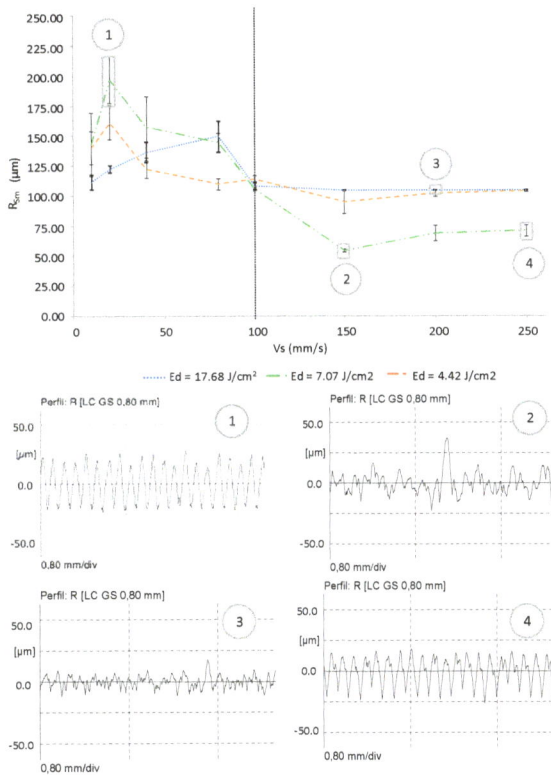

Figure 2. Roughness R_{Sm} behavior as a function of V_s.

3.2. Wetting Behavior of the Textured Surfaces

In this research, the ability to maintain control of the water retention proved to be a relevant aspect in order to improve the cooling and lubricating conditions for friction and sliding applications. Wettability behavior was investigated for surface topographies of various roughness features. Consequently, a control of the wetting behavior of the specimen was achieved through the variation of the laser processing parameters (E_d, V_s).

Under these considerations, an important dependency was detected between contact angle measures and V_s for textured surfaces, Figure 3, taking as starting point a value of 46.3° in the contact angle of the untreated surface. Higher intensity treatments developed under higher E_d and lower V_s resulted in a non-uniform asperity topography. This effect gave rise to an irregular distribution of the contact points on the surface. It also favored the breakage of surface tension between solid and liquid phases. Under these conditions, the absorption of water was promoted and the surface showed a better hydrophilic behavior than the initial alloy.

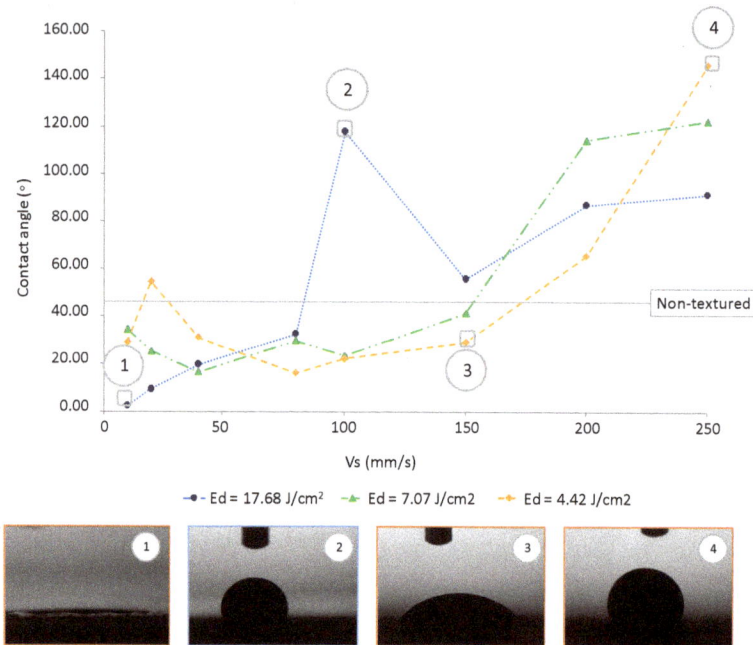

Figure 3. Contact angle as a function of V_s.

The existence of a singular point was detected under $E_d = 17.68$ J cm^{-2} and $V_s = 100$ mm s^{-1} conditions where a significant increase of the contact angle can be observed. It seems to be a relation between the development of texturized grooves with specific dimensions and geometry, and the generation of a uniform and distributed contact between phases.

When V_s was increased, an important growth occurred on the contact angle values, raising the hydrophobic behavior trends of the modified surface. This is especially caused by an increase in the compactness of the treated layer as a result of the development of textures with a higher contact area between different phases [19]. This phenomenon is reproduced under all the range of E_d analyzed in this work.

3.3. Friction Coefficient of the Tested Surfaces under Wet Conditions

Tribological wear behavior under wet conditions should be highly influenced by texturing process parameters as a consequence of wettability behavior. For this reason, hydrophilic surfaces should improve the sliding and cooling features of the contact area.

The evaluation of the textured surfaces by a pin on flat reciprocating tests confirmed a direct dependency between the laser modification process and friction coefficient (μ), due to its effects on roughness and wettability parameters. Friction coefficient values seemed to get stable from $V_s = 100$ mm s^{-1} to higher speeds.

On one hand, this behavior is coincident with R_{Sm} tendency confirming that the distance between texturing grooves influence on the friction and sliding process.

On the other hand, described stability trend is also related to significant growth of the contact angle for V_s higher than 100 mm s^{-1}. For these cases, the ability to absorb water was not enough to modify the sliding conditions compared with dry tests. This consideration showed similar behaviors for all the evaluated textures, as shown in Figure 4.

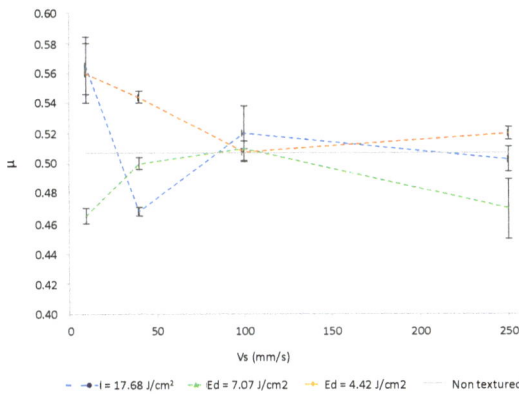

Figure 4. Friction coefficient behavior as a function of V_s.

The change of the textured surface as a function of sliding time was analyzed in Figure 5, with the aim to evaluate the durability of the treated surfaces under contact conditions. The lowest V_s was associated to the highest thickness layer and it was taken as a reference. The behaviors of μ values were analyzed during a test with a duration upper to 2000 s. Growth of the μ values was identified for non-textured samples, as a result of a progressive increase in the wear effects. Although, laser processed samples maintained a more stable behavior of μ values during all the test.

Because of this, uniform values of μ can be observed for the texturized samples in the tribological tests, confirming that this type of samples was not significantly affected by the wear phenomena. This effect is mainly due to the variations in the water absorption properties of the surface, as well as the improvement on the mechanical properties as hardness of the treated layer. However, an increase of the μ values was detected for the tests with $E_d = 17.68$ J cm^{-2} and $E_d = 4.42$ J cm^{-2} textured conditions.

After 1700 s, for low energy density of pulse treatments, an increase in the adhesive effects of the contact pair may appear. This effect results in the generation of a new contact pair with the same material, reducing the friction coefficient. In the case of 20 KHz, the thickness of the modified layer is higher than the other treatments increasing the protective effect of the treated area.

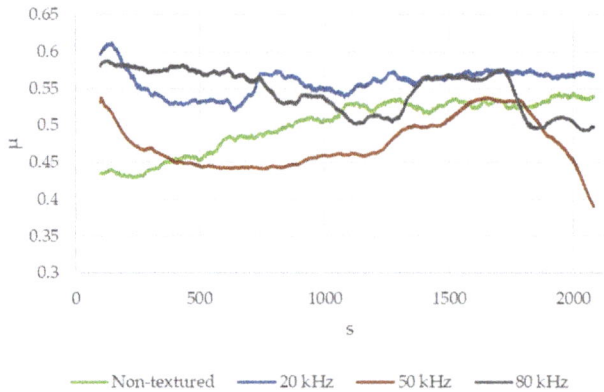

Figure 5. Friction coefficient evolution as a function of sliding time (V_s = 10 mm/s).

3.4. Wear Behavior of Modified Surfaces

Wear volume measures were significantly reduced in every textured sample with respect to non-textured conditions, Figure 6. Wear volume is not just affected by the wettability of the texture, but also by V_s parameter. This control parameter was related to the thermochemical process that produced the oxidation of the surfaces, the possible changes produced on the microstructure of the layer and the thickness of the affected layer [15,19,36]. This fact confirmed that laser surface processing results in a protective mechanism of the Ti6Al4V against tribological wear effects.

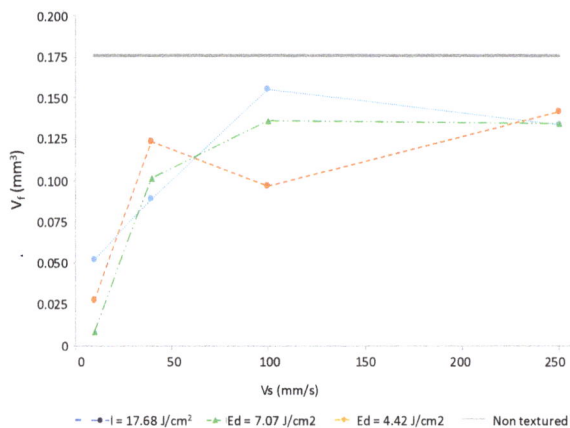

Figure 6. Wear volume evolution as a function of V_s.

Wear track analysis allowed to detect an increase of the reversal points of the sliding grove, raising the aggressive effects of the wear on both ends of the track. Over these areas, the carbide sphere experiment acceleration and breaking stages resulted in relevant friction and wear phenomena. In this sense, non-uniform wear tracks were observed for non-textured samples. This fact may develop irregular contact between the carbide sphere and alloy surface, causing an increase of the material worn volume, Figure 7.

For textured surfaces (E_d = 17.68 J cm^{-2}) the protective effect of the textured surface gives additional uniformity to the wear track, reducing the effects of the carbide balls at the end of the stroke.

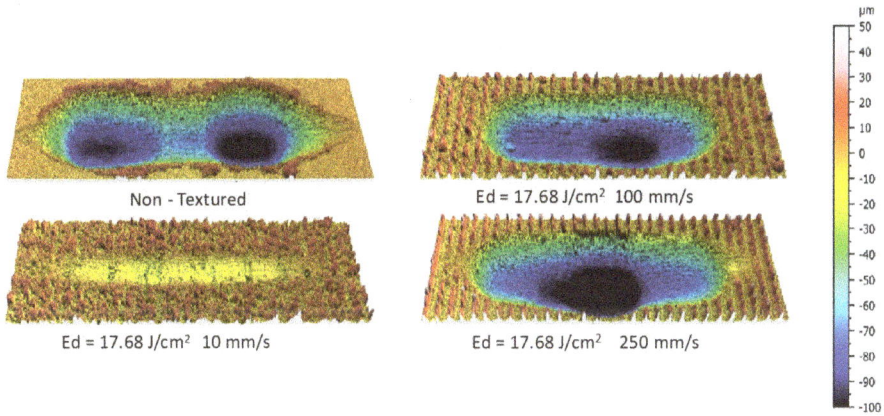

Figure 7. Worn morphologies of sliding track as a function of V_s.

3.5. Wear Effects on the Modified Surfaces

Friction and sliding phenomena over modified surfaces may induce the appearance of different wear mechanisms. The wear behavior of the tribological pair may be conditioned, specially based on the nature of modified surfaces, by modifications of the laser processing parameters.

Taking as starting point the wear effects of non-texturized surfaces, significant abrasive phenomenon was revealed in the bottom of the sliding grooves. This mechanism was mainly characterized by the appearance of scratches and embedded particles over the contact surface. As can be seen on Figure 6, detached particles from the Ti6Al4V surface subjected to cold-working effects became harder than initial alloy, giving rise to three-body abrasion mechanisms. Additionally, material subjected to plastic deformation was found on the external borders of the wear track, as can be seen in Figures 8 and 9.

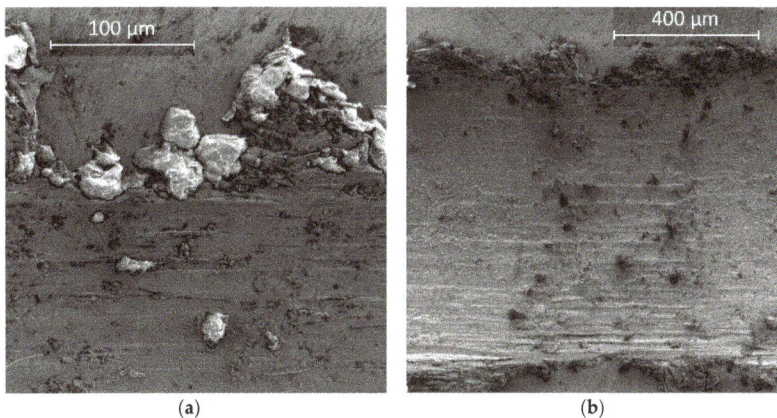

Figure 8. SEM of the non-textured wear track of Ti6Al4V sample; (**a**) 750×, (**b**) 200×.

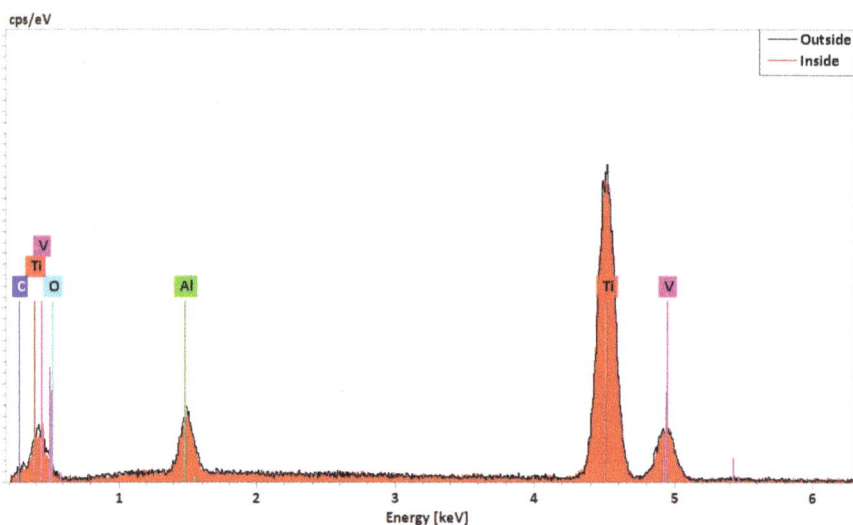

Figure 9. Energy dispersive X-ray spectroscopy (EDX) analysis of the wear track of the non-textured Ti6Al4V sample.

As detailed in a series of preliminary studies [19,36,37,42] the texturized process carried out under air atmosphere results in the appearance of oxidative phenomena. The incidence of oxidation process may induce the increase of hardness of the modified layer, changing the wear effects and the involved mechanisms. With the aim to analyze the influence of the laser treatment under wear conditions behavior of the surfaces, the texturized specimens using E_d = 7.07 J cm^{-2} were chosen as a reference for the evaluation of the V_s effects on the sliding and friction processes.

In this sense, a relevant increase of the thickness of the modified layer can be observed for lower V_s ranges, resulting in variations of the initial surface properties of the alloy. An important reduction of the track width was observed for V_s = 10 mm s^{-1}. This consideration was related to the decrease in the worn material volume which was mainly produced by the lowest contact angle between the water droplets and the surface used to evaluate the wetting behavior. This fact confirmed that texturized procedures with enough intensity significantly improve water retention properties of the surface giving rise to a decrease of wear effects.

Functional performance of the modified surfaces was affected by the increase of V_s values. Higher V_s implied the development of thinner layers. As a consequence, an important variation of the wear track width is shown in Figure 10, related to Figure 8 of the untreated specimen. In this situation, abrasive and adhesive mechanisms were observed on the bottom of the sliding groove, Figure 11.

Some extracted particles from the modified layer were harder than the initial alloy mainly due to oxidation. They were placed on the sliding way and caused scratches by the three body abrasion phenomena. Furthermore, the decrease of the contact angle as a function of V_s favored a hydrophobic trend of the textures and the lack of cooling effect provided by water. This fact may be the main cause of temperature increase in the contact area, resulting in the appearance of adhesive mechanism over the wear track and the carbide sphere. The cyclic nature of the adhesive mechanism caused the detachment of WC–Co sphere particles, as can be noticed in the EDX analysis performed on the sliding track, Figures 12 and 13.

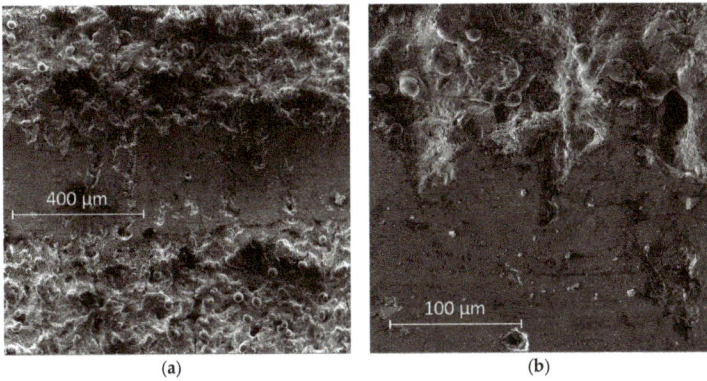

(a)   (b)

Figure 10. SEM of the wear track of the 50 Hz and 10 mm s^{-1} treated sample; (**a**) 200×, (**b**) 750×.

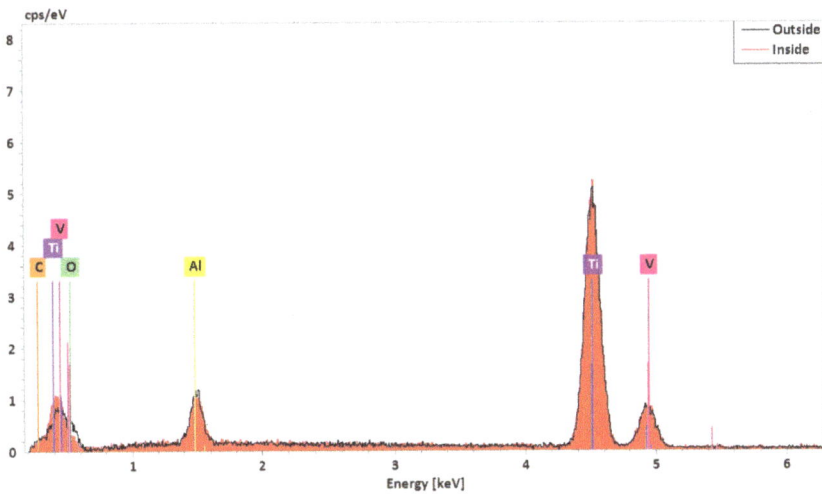

Figure 11. EDX analysis of the wear track of the 50 Hz and 10 mm s^{-1} treated sample.

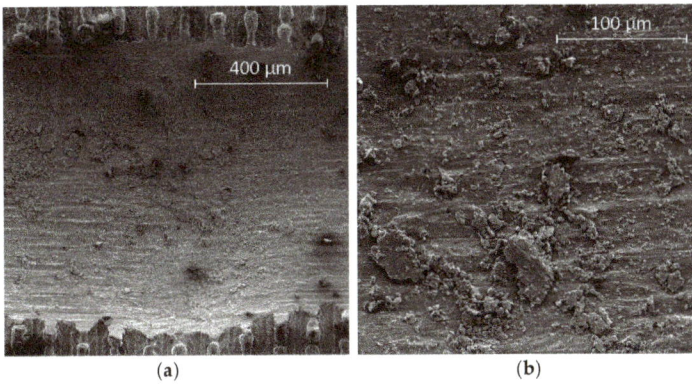

(a)   (b)

Figure 12. SEM of the wear track of the 50 Hz and 100 mm s^{-1} treated sample; (**a**) 200×, (**b**) 750×.

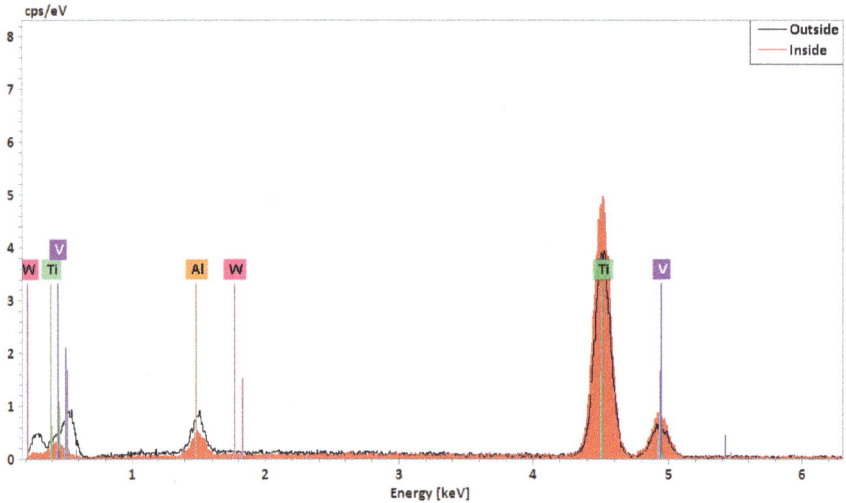

Figure 13. Wear track of the 50 Hz and 100 m s^{-1} sample EDX analysis.

Regarding carbide spheres, the adhesive mechanism was identified as the main factor affecting friction throughout the tribological tests. Because of this, adhesive behavior of the sphere surfaces is specially related to material worn volume, showing a relevant interest relation between the decrease of wear volume and sliding track width described in previous sections with respect to the adhered area of material on the sphere, Figure 14. This fact can be observed in Figure 15 where a configuration of a laser processing parameter with $V_s = 10$ mm s^{-1} resulted in smaller areas of increased adhesion on the surface of the carbide sphere, used as pins. Additionally, the increase of V_s is associated to the lack of wetting properties, as well as to the increase of wear volume of the textures, giving rise to higher adhered areas over the pins.

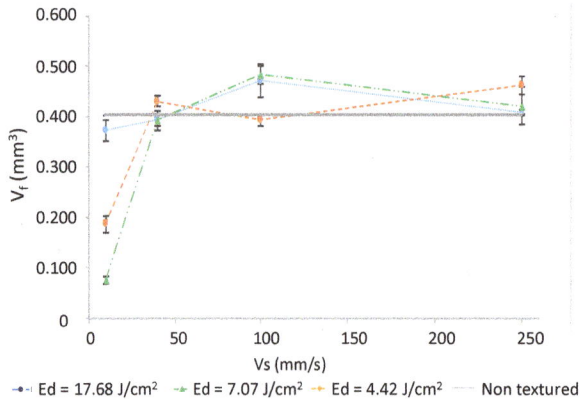

Figure 14. Worn material behavior as a function of V_s.

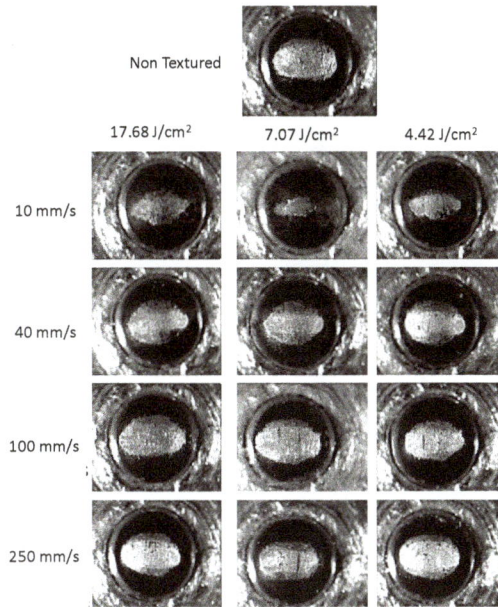

Figure 15. Adhesive wear effects on carbide spheres.

3.6. Comparative Evaluation between Wet and Dry Tests

The use of water as a base for lubricant and cooling fluids used in sliding and friction applications may improve the wear behavior of the surfaces. If the wettability properties of the contact area were adapted for higher water retention conditions through the development of micro- channels by the texturing process, the decrease of wear effects was more significant. V_s (10 mm s^{-1}) was taken as a reference in the evaluation of friction coefficient of reciprocating tests under wet and dry conditions. It was verified similar trends for the complete sliding range. Friction coefficient under dry and wet conditions is represented in Figure 16. It can be observed that friction coefficient decreased up to 35%, due to the wettability behavior of the sample.

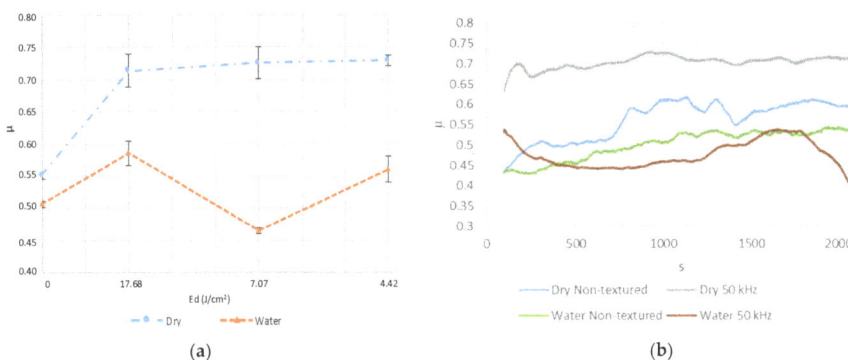

Figure 16. (**a**) Friction coefficient as a function of E_d in V_s = 10 mm s^{-1} for wet and dry conditions tests; (**b**) Friction coefficient as a function of time for wet and dry conditions tests for E_d = 7.07 J cm^{-2} and V_s = 10 mm s^{-1} laser treatment.

As can be observed in Figure 17, a decrease of the wear track dimensions regarding to a dry test was detected studying the wear effects under wet conditions of texturing surfaces (E_d = 7.07 J cm^{-2}; V_s = 10 mm s^{-1}). In this sense, the use of water may also reduce the volume of worn material in non-textured surfaces, however an adverse effect was observed. Wear track for non-textured surface under wet tribological tests showed a lack of uniformity, in terms of width and depth of the track. This fact could explain the unstable friction coefficient values obtained and shown in Figure 16.

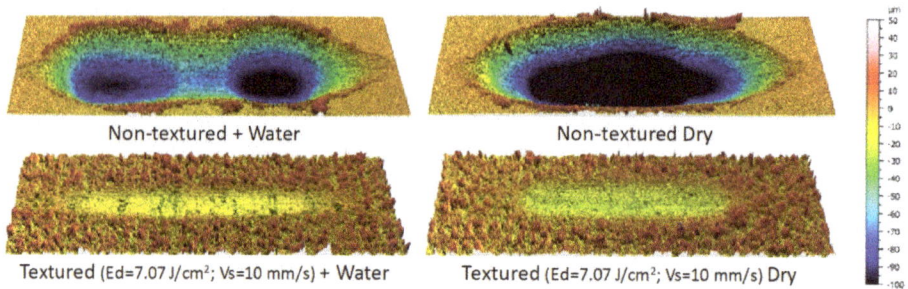

Figure 17. Effect of texturing treatment on wear track of dry and wet tribological tests.

The width and depth of the sliding track and the associated wear volume could be significantly reduced by the use of laser texturing treatments. This effect is mainly accomplished by the modification of contact area between the tribological pair elements, and the improvements on the surface properties of the alloy, such as hardness and wettability. Under this consideration, a protective function of the modified layer may be considered for friction and sliding applications.

4. Conclusions

Laser texturing treatments allow the development of special topographies on the surface of mechanical components. Dimensions and geometry of the texturing tracks can be controlled throughout the laser process, being V_s and E_d the main variables that govern the specific features of micro-textures. For instance, lower incidence of energy, induced by higher V_s, results in a decrease of the maximum size and geometrical characteristic of the textured grooves.

Micro-geometrical properties modify the wetting behavior of the surface. Contact angle measured values were increased up to a 300% compared to the non-textured surfaces for the highest E_d and scanner speed used. However, the specimens textured with the lower E_d and V_s produced a completely hydrophilic surface.

Even though, R_z values are increased by the laser treatment, friction coefficient values are about ±10% of the untreated friction coefficient under water conditions. Nevertheless, wear track volume is reduced by up to 70% by lower V_s treatments (<50 mm s^{-1}) concerning to untreated surfaces for all the analyzed ranges of E_d under both conditions, water and dry tests. Furthermore, galling and adhesion effects on the carbide balls are reduced up to more than 90%.

Author Contributions: Conceptualization, J.M.V.M., I.D.S.I. and P.I.V.; Methodology, J.M.V.M., I.D.S.I., P.I.V. and J.S.; Software, I.D.S.I.; Validation, J.M.V.M.; Formal J.M.V.M. and I.D.S.I.; Investigation, J.M.V.M. and J.S.; Resources, J.M.V.M., P.I.V. and J.S.; Writing-Original Draft Preparation, J.M.V.M., and I.D.S.I.; Writing-Review & Editing, P.I.V. and J.S.; Supervision, J.M.V.M. and J.S.; Project Administration, J.S.; Funding Acquisition, J.S.

Funding: This work has received financial support from the Spanish Government (MINECO/AEI/FEDER, No. DPI2017-84935-R).

Conflicts of Interest: The authors declare no conflict of interest.

References

1. Shahali, H.; Jaggessar, A.; Yarlagadda, P.K. Recent advances in manufacturing and surface modification of titanium orthopaedic applications. *Procedia Eng.* **2017**, *174*, 1067–1076. [CrossRef]
2. James, A.S.; Thomas, K.; Mann, P.; Wall, R. The role and impacts of surface engineering in environmental design. *Mater. Des.* **2005**, *26*, 594–601. [CrossRef]
3. Tian, Y.S.; Chen, C.Z.; Li, S.T.; Huo, Q.H. Research progress on laser surface modification of titanium alloys. *Appl. Surf. Sci.* **2005**, *242*, 177–184. [CrossRef]
4. Asri, R.I.M.; Harum, W.S.W.; Samykano, M.; Lah, N.A.C.; Ghani, S.A.C.; Tarlochan, F.; Raza, M.R. Corrosion and surface modification of biocompatible metals: A review. *Mater. Sci. Eng. C* **2017**, *77*, 1261–1274. [CrossRef] [PubMed]
5. Weng, F.; Chuanzhong, C.; Yu, H. Research status of laser cladding on titanium and its alloys: A review. *Mater. Des.* **2014**, *58*, 412–425. [CrossRef]
6. Biswas, A.; Chatterjee, U.K.; Li, L.; Manna, I.; Majumdar, J.D. Laser assisted surface modification of Ti-6Al-4V for bioimplant application. *Surf. Rev. Lett.* **2007**, *14*, 531–534. [CrossRef]
7. Mohammed, M.T.; Khan, Z.A.; Siddiquee, A.N. Surface modification of titanium materials for developing corrosion behavior in human body environment: A review. *Procedia Mater. Sci.* **2014**, *6*, 1610–1618. [CrossRef]
8. Wang, D.; Wang, Y.; Wu, S.; Lin, H.; Yang, Y.; Fan, S.; Gu, C.; Wang, J.; Song, C. Customized a Ti6Al4V bone plate for complex pelvic fracture by selective laser melting. *Materials* **2017**, *10*, 35. [CrossRef]
9. Chan, C.-W.; Lee, S.; Smith, G.; Sarri, G.; Ng, C.-H.; Sharba, A.; Man, H.-C. Enhancement of wear and corrosion resistance of beta titanium alloy by laser gas alloying with nitrogen. *Appl. Surf. Sci.* **2016**, *367*, 80–90. [CrossRef]
10. Attar, H.; Ehthemam-Haghighi, S.; Kent, D.; Okulov, I.V.; Wendrock, H.; Bönisch, M.; Volegov, A.S.; Calin, M.; Eckert, J.; Dargusch, M.S. Nanoindentation and wear properties of Ti and Ti-TiB composite materials produced by selective laser melting. *Mater. Sci. Eng. A* **2017**, *688*, 20–26. [CrossRef]
11. Sebastian, D.; Yao, C.; Lian, I. Mechanical durability of engineered superhydrophobic surfaces for anti-corrosion. *Coatings* **2018**, *8*, 162. [CrossRef]
12. Aniolek, K.; Kupka, M.; Barylski, A. Sliding wear resistance of oxide layers formed on a titanium surface during thermal oxidation. *Wear* **2016**, *356–357*, 23–29. [CrossRef]
13. May, A.; Agarwal, N.; Lee, J.; Lambert, M.; Akkan, C.K.; Nothdurft, F.P.; Aktas, O.C. Laser induced anisotropic wetting on Ti-6Al-4V surfaces. *Mater. Lett.* **2015**, *138*, 21–24. [CrossRef]
14. Akbarzadeh, A.; Khonsari, M.M. Effect of untampered plasma coating and surface texturing on friction and running-in behavior of piston rings. *Coatings* **2018**, *8*, 110. [CrossRef]
15. Cinca, N.; Cygan, S.; Senderowski, C.; Jaworska, L.; Dosta, S.; Cano, I.G.; Guilemany, J.M. Sliding wear behavior of Fe-Al coatings at high temperatures. *Coatings* **2018**, *8*, 268. [CrossRef]
16. Ahuir-Torres, J.I.; Arenas, M.A.; Perrie, W.; Dearden, G.; de Damborenea, J. Surface texturing of aluminium alloy AA2024-T3 by picosecond laser: Effect on wettability and corrosion properties. *Surf. Coat. Technol.* **2017**, *321*, 279–291. [CrossRef]
17. Matschegewski, C.; Staehlke, S.; Birkholz, H.; Lange, R.; Beck, U.; Engel, K.; Nebe, J.B. Automatic actin filament quantification of osteoblasts and their morphometric analysis on microtextured silicon-titanium arrays. *Materials* **2012**, *5*, 1176–1195. [CrossRef]
18. Demir, A.G.; Maressa, P.; Previtali, B. Fibre laser texturing for surface functionalization. *Phys. Procedia* **2013**, *41*, 759–768. [CrossRef]
19. Vazquez-Martinez, J.M.; Salguero Gomez, J.; Mayuet Ares, P.F.; Fernandez-Vidal, S.R.; Batista Ponce, M. Effects of laser microtexturing on the wetting behavior of Ti6Al4V alloy. *Coatings* **2018**, *8*, 145. [CrossRef]
20. Ancona, A.; Joshi, G.S.; Volpe, A.; Scaraggi, M.; Lugarà, P.M.; Carbone, G. Non-uniform laser surface texturing of an un-tapered square pad for tribological applications. *Lubricants* **2017**, *5*, 41. [CrossRef]
21. Ukar, E.; Lamikiz, A.; Martinez, S.; Arrizubieta, I. Laser texturing with conventional fiber laser. *Procedia Eng.* **2015**, *132*, 663–670. [CrossRef]
22. Bormashenko, E. Progress in understanding wetting transitions on rough surfaces. *Adv. Colloid Interface Sci.* **2015**, *222*, 92–103. [CrossRef] [PubMed]
23. Otitoju, T.A.; Ahmad, A.L.; Ooi, B.S. Superhydrophilic (superwetting) surfaces: A review on fabrication and application. *J. Ind. Eng. Chem.* **2017**, *47*, 19–40. [CrossRef]

24. Belhadjamor, M.; Belghith, S.; Mezlini, S.; El Mansori, M. Effect of the surface texturing scale on the self-clean function: Correlation between mechanical response and wetting behavior. *Tribol. Int.* **2017**, *111*, 91–99. [CrossRef]

25. Wojciechowski, L.; Kubiak, K.J.; Mathia, T.G. Roughness and wettability of surfaces in boundary lubricated scuffing wear. *Tribol. Int. B* **2016**, *93*, 593–601. [CrossRef]

26. Liang, Y.; Shu, L.; Natsu, W.; He, F. Anisotropic wetting characteristics versus roughness on machined surfaces of hydrophilic and hydrophobic materials. *Appl. Surf. Sci.* **2015**, *331*, 41–49. [CrossRef]

27. Attar, H.; Calin, M.; Zhang, L.C.; Scudino, S.; Eckert, J. Manufacture by selective laser melting and mechanical behavior of commercially pure titanium. *Mater. Sci. Eng. A* **2014**, *593*, 170–177. [CrossRef]

28. Ali, N.; Bashir, S.; Kalsoom, U.; Akram, M.; Mahmood, K. Effect of dry and wet ambient environment on the pulsed laser ablation of titanium. *Appl. Surf. Sci.* **2013**, *270*, 49–57. [CrossRef]

29. Oh, J.M.; Lee, B.G.; Cho, S.W.; Choi, G.S.; Lim, J.W. Oxygen effects on the mechanical properties and lattice strain of Ti and Ti-6Al-4V. *Met. Mater. Int.* **2011**, *17*, 733–736. [CrossRef]

30. Lavisse, L.; Jouvard, J.M.; Imhoff, L.; Heintz, O.; Korntheuer, J.; Langlade, C.; Bourgeois, S.; Marco de Lucas, M.C. Pulsed laser growth and characterization of thin films on titanium substrates. *Appl. Surf. Sci.* **2007**, *253*, 8226–8230. [CrossRef]

31. Fasai, A.Y.; Mwenifumbo, S.; Rahbar, N.; Chen, J.; Li, M.; Beye, A.C. Nano-second UV laser processed micro-grooves on Ti6Al4V for biomedical applications. *Mater. Sci. Eng. C* **2009**, *29*, 5–13. [CrossRef]

32. Xing, Y.; Deng, J.; Gao, P.; Gao, J.; Wu, Z. Angle-dependent tribological properties of AlCrN coatings with microtextures induced by nanosecond laser under dry friction. *Appl. Phys. A* **2018**, *124*, 294. [CrossRef]

33. Bonse, J.; Koter, R.; Hartelt, M.; Spaltmann, D.; Pentzien, S.; Höhm, S.; Rosenfeld, A.; Krüger, J. Tribological performance of femtosecond laser-induced periodic surface structures on titanium and a high toughness bearing steel. *Appl. Surf. Sci.* **2015**, *336*, 21–27. [CrossRef]

34. Gnilitskyi, I.; Rotundo, F.; Martini, C.; Pavlov, I.; Ilday, S.; Vovk, E.; Ilday, F.O.; Orazi, L. Nano patterning of AISI 316L stainless steel with Nonlinear Laser Lithography: Sliding under dry and oil-lubricated conditions. *Tribol. Int.* **2016**, *99*, 67–76. [CrossRef]

35. Mahamood, R.M.; Akinlabi, E.T.; Shukla, M.; Pityana, S. Scanning velocity influence on microstructure, microhardness and wear resistance of laser deposited Ti6Al4V/TIC composite. *Mater. Des.* **2013**, *50*, 656–666. [CrossRef]

36. Vrancken, B.; Thijs, L.; Kruth, J.P.; Van Humbeeck, J. Heat treatment of Ti6Al4V produced by selective laser melting: Microstructure and mechanical properties. *J. Alloys Compd.* **2012**, *541*, 177–185. [CrossRef]

37. Vázquez Martínez, J.M.; Salguero Gómez, J.; Batista Ponce, M.; Botana Pedemonte, F.J. Effects of laser processing parameters on texturized layer development and surface features of Ti6Al4V alloy samples. *Coatings* **2018**, *8*, 6. [CrossRef]

38. Zhao, Y.; Du, H. Effect of laser scanning speed on the wear behavior of nano-SiC-modified Fe/WC composite coatings by laser remelting. *Coatings* **2018**, *8*, 241. [CrossRef]

39. Khana, R.; Ong, J.L.; Oral, E.; Narayan, R.J. Progress in wear resistant materials for total hip arthroplasty. *Coatings* **2017**, *7*, 99. [CrossRef]

40. Veiga, C.; Davim, J.P.; Loureiro, A.J.R. Properties and applications of titanium alloys: A brief review. *Rev. Adv. Mater. Sci.* **2012**, *32*, 133–148.

41. Leuders, S.; Thone, M.; Riemer, A.; Niendorf, T.; Troster, T.; Richard, H.A.; Maier, H.J. On the mechanical behavior of titanium alloy Ti6Al4V manufactured by selective laser melting: Fatigue resistance and crack growth performance. *Int. J. Fatigue* **2013**, *48*, 300–307. [CrossRef]

42. Patel, D.S.; Singh, A.; Balani, K.; Ramkumar, J. Topographical effects of laser surface texturing on various time-dependent wetting regimes in Ti6Al4V. *Surf. Coat. Technol.* **2018**, *349*, 816–829. [CrossRef]

43. Chen, L.; Liu, Z.; Shen, Q. Enhancing tribological performance by anodizing micro-textured surfaces with nano-MoS$_2$ coatings prepared on aluminum-silicon alloys. *Tribol. Int.* **2018**, *122*, 84–95. [CrossRef]

44. Vlădescu, S.-C.; Olver, A.; Pegg, I.; Reddyhoff, T. The effects of surface texture in reciprocating contacts—An experimental study. *Tribol. Int.* **2015**, *82*, 28–42. [CrossRef]

45. AlMangour, B.; Grzesiak, D.; Cheng, J.; Ertas, Y. Thermal behavior of the molten pool, microstructural evolution, and tribological performance during selective laser melting of TiC/316L stainless steel nanocomposites: Experimental and simulation methods. *J. Mater. Process. Technol.* **2018**, *257*, 288–301. [CrossRef]

46. Varenberg, M.; Halperin, G.; Etsion, I. Different aspects of the role of wear debris in fretting wear. *Wear* **2002**, *252*, 902–910. [CrossRef]

47. *ISO 4287:1997 Geometrical Product Specifications (GPS)—Surface Texture: Profile Method—Terms, Definitions and Surface Texture Parameters*; International Organization Standardization (ISO): Geneva, Switzerland, 1997.

48. Qu, J.; Truhan, J.J. An efficient method for accurately determining wear volumes of sliders with non-flat wear scars and compound curvatures. *Wear* **2006**, *261*, 848–855. [CrossRef]

49. *ASTM G133-05 Standard Test Method for Linearly Reciprocating Ball-on-Flat Sliding Wear*; ASTM Standard: West Conshohocken, PA, USA, 2016.

Article

Cracking, Microstructure and Tribological Properties of Laser Formed and Remelted K417G Ni-Based Superalloy

Shuai Liu [1], Haixin Yu [1], Yang Wang [1], Xue Zhang [1], Jinguo Li [2], Suiyuan Chen [1] and Changsheng Liu [1,*]

[1] Key Laboratory for Anisotropy and Texture of Materials Ministry of Education, School of Materials Science and Engineering, Northeastern University, Shenyang 110819, China; 13840524163@163.com (S.L.); 13840362963@163.com (H.Y.); 15040369625@163.com (Y.W.); 18804025352@163.com (X.Z.); chensy@atm.neu.edu.cn (S.C.)

[2] Institute of Metal Research, Chinese Academy of Sciences, Shenyang 110006, China; jgli@imr.ac.cn

* Correspondence: csliu@mail.neu.edu.cn; Tel.: +86-24-83691579

Received: 15 November 2018; Accepted: 21 January 2019; Published: 24 January 2019

Abstract: The K417G Ni-based superalloy is widely used in aeroengine turbine blades for its excellent properties. However, the turbine blade root with fir tree geometry experiences early failure frequently, because of the wear problems occurring in the working process. Laser forming repairing (LFR) is a promising technique to repair these damaged blades. Unfortunately, the laser formed Ni-based superalloys with high content of (Al + Ti) have a high cracking sensitivity. In this paper, the crack characterization of the laser forming repaired (LFRed) K417G—the microstructure, microhardness, and tribological properties of the coating before and after laser remelting—is presented. The results show that the microstructure of as-deposited K417G consists of γ phase, γ' precipitated phase, $\gamma + \gamma'$ eutectic, and carbide. Cracking mechanisms including solidification cracking, liquation cracking, and ductility dip cracking are proposed based on the composition of K417G and processing characteristics to explain the cracking behavior of the K417G superalloy during LFR. After laser remelting, the microstructure of the coating was refined, and the microhardness and tribological properties was improved. Laser remelting can decrease the size of the cracks in the LFRed K417G, but not the number of cracks. Therefore, laser remelting can be applied as an effective method for strengthening coatings and as an auxiliary method for controlling cracking.

Keywords: K417G Ni-based superalloy; laser forming repairing; laser remelting; microstructure; cracking behavior; tribology

1. Introduction

Because of the objective reality of enhancing aeroengine performance, blades must work in an environment of high temperature, overloading, and high frequency vibration [1]. The Ni-based superalloy K417G containing a high content of Al + Ti (>7.0 wt.%) is widely used in aeroengine turbine blades for its excellent high-temperature properties and relatively low fabricating cost [2–5]. However, the turbine blade root with fir tree geometry experiences early failure frequently, because of the wear problems occurring in the working process [1]. From an efficiency and economic point of view, it is more appealing and significant to repair the defected or damaged blades instead of replacing them with new ones. Laser forming repairing (LFR), also called laser cladding, is a kind of metal additive manufacturing technology. It can be applied to form a repaired coating that recovers complex or various defected parts up to certain degree and to form a metallurgical bond between substrate and coating, without degrading the inherent service properties of the parts [6–8].

Unfortunately, cracking behavior frequently occurs when laser rapid forming technology is used to manufacture nickel-based superalloy containing a high content of Al + Ti (>7.0 wt.%), which is the most harmful defect that seriously affects the reliability of an aeroengine. Consequently, it is significant to explore the cracking mechanism of the nickel-based superalloy and to seek methods of controlling the cracking behavior. Ojo and Chaturvedi deemed that the constitutional liquation of γ' phase was the main factor in liquation cracks in the Inconel 738 welding process [9,10]. Similarly, Li et al. repaired the damaged K465 superalloy turbine blades by LFR and came to the conclusion that the constitutional liquation of γ' phase resulted in the formation of liquation films during the K465 repairing process [11]. Tancret et al. computed the liquated γ' phase temperature of Inconel 738 by Thermo-Calc software and analyzed the relationship between the heating rate and the dissolution behavior of γ' phase [12]. Zhou and Li et al. pointed out that low melting point phases at grain boundaries such as $\gamma + \gamma'$ eutectics and carbides were the main factors in the cracking behavior in the K3 nickel-based superalloy during laser cladding [13,14]. Yang et al. proposed three cracking mechanisms based on the composition of Rene 104 and processing characteristics to explain the cracking behavior of the Rene 104 superalloy during direct laser fabrication [15].

Although a few published papers have investigated different mechanisms of nickel-based superalloys during the laser forming process, limited works have been carried out on the K417G superalloy. So far, the microstructure and properties of laser formed K417G, the cracking mechanism, and control methods of controlling the cracking behavior are still unclear. Moreover, except for component factors and processing parameters, another approach to affect the cracking behavior is post treatment. Laser remelting is considered as an effective post treatment to improve the quality and properties of coatings. It has been extensively adopted to prepare coatings with a dense structure and excellent properties [16–20]. Therefore, laser remelting can be attempted to decrease or even eliminate the cracks in the laser forming repaired (LFRed) K417G. In this paper, the microstructure observation and crack analysis of the LFRed K417G are presented. Then, cracking mechanisms are proposed, taking the chemical composition of the K417G superalloy and laser processing characteristics of LFR into account. Finally, the effects on microstructure, cracking behavior, hardness, and tribological properties of the LFRed K417G after laser remelting are investigated.

2. Materials and Methods

The substrate used in this experiment was the as-cast K417G superalloy with dimensions of ø 25 mm × 8 mm. The K417G spherical powder was supplied by Institute of Metal Research of Chinese Academy of Sciences (Shenyang, China) and refined by the gas atomization method. The particle size of powder was 50–150 μm and its specific elemental composition was 0.14C, 9.84Cr, 6.37Al, 4.79Ti, 11.4Co, 3.18Mo, 2.80Fe, and balance Ni (wt.%). The laser equipment used in this experiment was a laser direct deposition forming system (Key Laboratory for Anisotropy and Texture of Materials, Ministry of Education, Shenyang, China), which mainly consisted of a YAG-1000W fiber laser, protective atmosphere device, self-designed coaxial ring powder feeding device, circulating cooling device, and computer system for forming control. In the laser repairing process, the positive defocusing modes were adopted to obtain a smaller dilution ratio and a higher cladding efficiency, the defocus amount was 4 mm, and the spot diameter was 1.8 mm. High purity argon (99.99%) was used as the bath protection gas to prevent oxidation, and the shielding gas flow rate was 7 L/min. According to the relevant studies [21–23], the process parameters of LFR were optimized. The LFR process parameters of each layer are shown in Table 1 and the schematic of laser scanning path is shown in Figure 1. The deposited coating was ten layers in order to ensure adequate thickness. The laser remelting process parameters were the same as the deposition process, except that no powder was fed. When the deposition of each layer was completed, the powder feeding was stopped and the surface of deposited layer was laser scanned again along the original path at that height. Each layer was remelted once. Then, the focus was raised by 0.4 mm to continue the process of deposition and remelting of the next layer. Because there were ten layers of the coating, the total of remelting times was ten. The remelting process did not bring about a great change in thickness of each deposited layer. Before and after remelting, the thickness of each layer was about 0.4 mm. The thicknesses of two ten-layer coatings were similar, both of which were 3.8 mm.

Table 1. Main process parameters of laser forming repairing.

Laser Power (W)	Scanning Speed (mm/s)	Powder Feeding Amount (g/min)	Powder Flow (L/min)	Overlap Rate (%)	Z-Axis Lift (mm)	Interlayer Cooling Time (min)
600	5.4	5	3.5	40	0.4	10

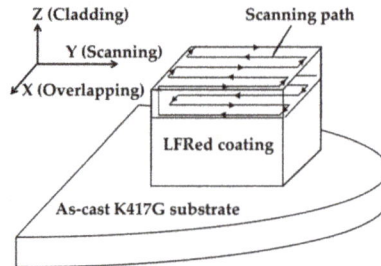

Figure 1. The schematic of laser scanning path. LFR—laser forming repairing.

After preparing, the LFRed coatings were cut into reasonable size blocks together with the substrate by numerically controlled wire-cutting. After being mechanically ground and polished, the sectioned samples were electrolytically etched in 15 g CrO_3 + 10 mL H_2SO_4 + 150 mL H_3PO_4 at 5 V for 35–50 s. Samples were ultrasonically cleaned after corrosion for 10 min, and finally rinsed with absolute ethanol and dried. The phase composition was measured using an X-ray diffractometer (XRD) (Smartlab-9000, Rigaku, Tokyo, Japan) machine and the main operating parameters including a 40 kV voltage, 250 mA current, Cu Kα radiation, 0.02° angle step-length, and 5°/min scanning rate. Scanning electron microscope (SEM) (JSM-6510A, JEOL, Tokyo, Japan) and its own energy spectrum analyzer (EDS) were used to observe the microstructure and analyze micro-area composition and wear surface of samples. The dendritic spacing and crack size were measured by Image-Pro Plus 6.0 image analysis software (Image-Pro Plus software, Media Cybernetics, Bethesda, MD, USA). The microhardness was measured by a digital micro vickers hardness tester (401MVD, Wolpert Wilson, Norwood, MA, USA). The test areas were regions from the top of the coating to the substrate in the longitudinal section of Y–Z. A load of 200 g was applied, and the holding time was 10 s. The microhardness of each sample was measured three times, and the average of the three results was taken as the microhardness of the sample. The tribological properties of samples were evaluated using Universal friction and wear tester (Nanovea, Irvine, CA, USA) and its own 3D contact surface profiler. The samples with dimensions of ø 15 mm × 10 mm were prepared to conduct the wear experiment at room temperature. The radius of the wear tracks was set to 3 mm using φ6 mm Si_3N_4 balls as a counterpart. The measurements were implemented for a sliding length of 54 m with a speed of 15 mm/s, a load of 10 N, and a relative humidity of 60% ± 3%.

3. Results and Discussion

3.1. Microstructure and Main Phases of LFRed K417G Superalloy

Samples were examined in order to determine the phase composition using an X-ray diffraction analyzer (XRD). Figure 2 is an XRD diffraction pattern of the LFRed K417G. It can be seen that the coating mainly contains γ solid solution, Al0.5CNi3Ti0.5 carbide, and γ'-Ni3(Al,Ti) strengthening precipitation phase. Figure 3 shows the SEM images and the EDS analysis results of line scanning on the X–Y section. The results indicate that Ti and Mo elements are significantly segregated between the dendrites, while Cr and Co are evenly distributed in the dendrites. Figure 4 shows the typical microstructure of the as-deposited K417G on the X–Z section and the X–Y section. The darker areas in Figure 4a are dendrites, while the brighter areas are interdendritic zones, the measured dendrite spacing is 10–18 μm. There are mainly three types of precipitates between dendrites, namely, finely distributed dot-like precipitates, white block-shaped precipitates, and gray-white fishbone-like tissue. The proportion of white block-shaped precipitates and gray-white fishbone-like tissue is higher, occupying half of the interdendritic zones and distributing nonuniformly.

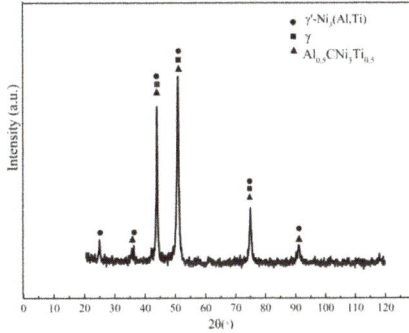

Figure 2. X-ray diffractometer (XRD) diffraction pattern of the laser forming repaired (LFRed) K417G.

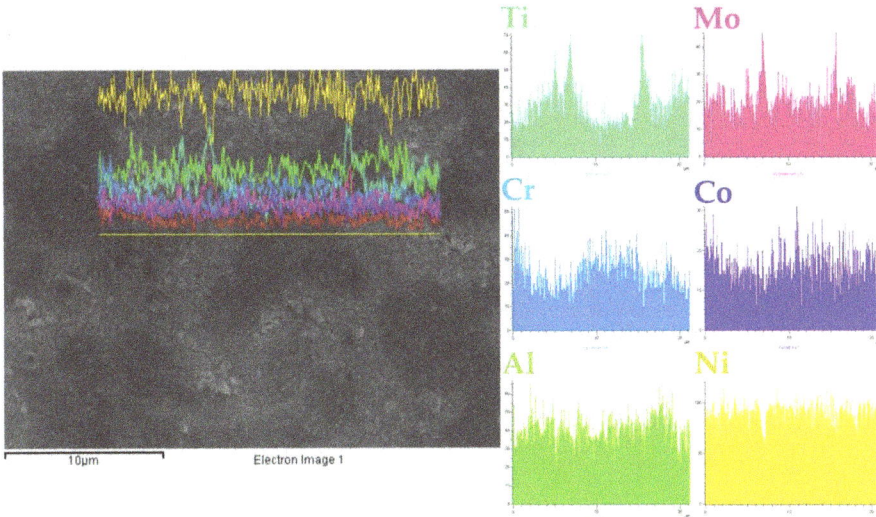

Figure 3. Energy dispersive spectrometer (EDS) results of line scanning on the X–Y section.

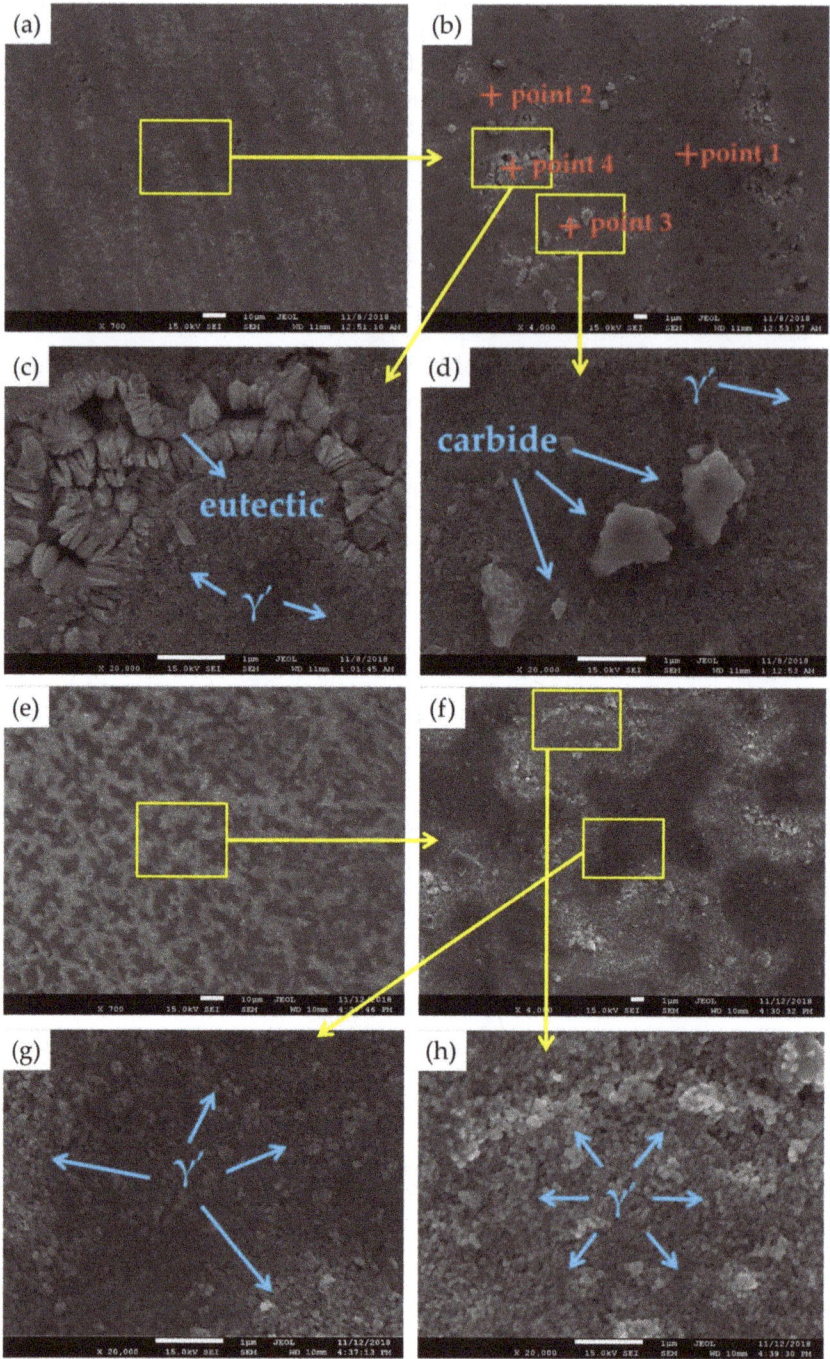

Figure 4. Typical microstructure on (**a–d**) the X–Z section and (**e–h**) the X–Y section of the as-deposited K417G.

In order to further analyze the microstructure and phase composition, energy dispersive spectrometer (EDS) of point scanning is performed on typical locations on the microstructure. As shown in Table 2, the results indicate that the composition of the dark-gray zone in dendrites (point 1 in Figure 4b) is similar to that of the original powder, while the contents of Al, Ti, and Mo are lower than the original powder. This is because during the solidification process, because the laser deposition is a near-rapid cooling process, the Al, Ti, and Mo element will be partly segregated in the remaining liquid phase during the non-equilibrium solidification process. As can be seen from Figure 4g, there are also dot-like precipitates in the dark-gray zone, which are the γ' phases precipitated on the matrix. Therefore, the dark-gray zone (point 1 in Figure 4b) is a two phase γ plus γ' microstructure. The element distribution of finely distributed dot-like precipitates between the dendrites (point 2 in Figure 4b) is close to that of the matrix, but the content of Al and Ti is increased. Combined with the previous XRD results and morphological comparison of precipitates in some literature about Ni-based superalloys [24,25], it can be concluded that the location of point 2 is also γ phase plus γ' phase Ni3(Al,Ti); in addition, the aggregation of Ti will make the γ' phase coarse. Therefore, the morphology and content of γ' phase between the dendrites are relatively different from those in the dendrites, as we can see from Figure 4g,h. The location of point 3 (in Figure 4b) shows the white block-shaped precipitates distributed between dendrites. An elemental analysis shows that the C, Ti, and Mo elements are enriched in the particle precipitated phase compared with the matrix phase. Its morphology is characterized by MC-type carbides, where M represents the metal elements Al, Ti, and Mo. According to previous XRD analysis, it is presumed that the precipitation form of MC should be (Al, Ti, Mo) C. However, considering some types of carbides such as M6C, M23C6, and so on are usually hard to detect in XRD, it is difficult to determine whether other types of carbides are present in as-deposited K417G [25,26]. The content of elemental composition at point 4 is similar to that at point 2. The contents of C and Ti are slightly higher, and the form is clustered or fishbone. This is $\gamma + \gamma'$ eutectic structure. The matrix is first formed during solidification, Al and Ti elements are segregated in the liquid to precipitate and grow the γ' phase in advance, and finally the remaining liquid is pushed to the interdendritic position to cause eutectic reaction at a lower temperature; then, the eutectic structure forms [25,27–29].

Table 2. Elements' concentration of different test points (mass fraction %).

Element	C	Al	Ti	Cr	Fe	Ni	Mo	Co
Point 1	1.79	4.65	1.91	9.33	7.87	62.34	2.81	9.45
Point 2	2.61	5.85	7.36	5.73	2.46	64.13	4.62	3.52
Point 3	8.72	5.56	6.24	9.23	5.62	51.23	3.20	5.72
Point 4	3.87	3.52	6.03	7.52	3.16	61.25	3.23	5.26
Powders	0.14	6.37	4.79	9.84	2.80	61.2	3.18	11.4

3.2. Cracking Behavior and Mechanisms of LFRed K417G Superalloy

3.2.1. Crack Observation and Analysis

Unfortunately, severe cracking behavior occurs in the as-deposited K417G samples. Figure 5a shows the cracks on the X–Z section, the existence of cracks with the length in the range of 0.2 to 2 mm in the repair zone (RZ) and the heat affected zone (HAZ) can be found. In addition, the cracks on the X–Z section are approximately parallel to the depositional direction, and have the same orientation as the columnar crystal. This is because the grain boundary is considered as a favorable crack initiation site, so that cracks tend to form along the grain boundary, except for a few initial points in the pores, which is similar to the results presented by Carter et al. [30]. The crack count and crack density, defined as the total number of cracks per unit area by Ghosh and Partha, are measured to estimate the cracking sensitivity [31]. Figure 5b gives the crack count and density of the as-deposited K417G samples. According to the observation of different sections and the statistical results in Figure 5, it can be concluded that compared with most superalloys, the cracking sensitivity of the as-deposited K417G is more serious [9–15,32].

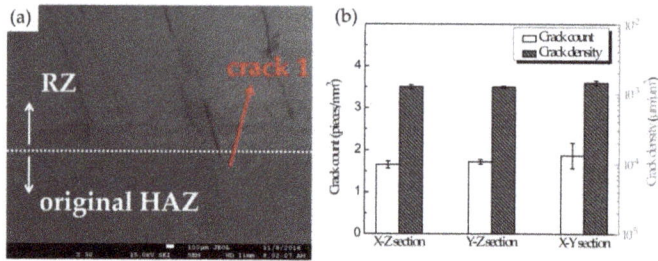

Figure 5. (**a**) The cracks on the X–Z section and (**b**) the crack count and density on three sections of as-deposited K417G. RZ—repair zone; HAZ—heat affected zone.

This characteristic of cracking behavior is mainly influenced by the composition of K417G and the processing of LFR. Firstly, K417G has a very high content of Al and Ti (11 wt.%), which are γ'-Ni3(Al,Ti) forming elements. Consequently, a large amount of γ' phase exists in the K417G superalloy. On one hand, these γ' phases play a role in strengthening the material. On the other hand, the contraction stresses during the precipitation of γ'-Ni3(Al,Ti) and the forming of low melting point $(\gamma + \gamma')$ eutectics will be caused by these γ' phases, which will dramatically increase the cracking sensitivity of the alloy [25,33]. Secondly, the processing characteristic of LFR is another vital factor that affects the cracking sensitivity. LFR is a process of layer-by-layer superposition. As shown in Figure 6a, the LFRed K417G has an obvious layer-by-layer structure. Also, deposited layers of different heights undergo different thermal cycles. Five types of the thermal cycles with different peak temperatures have been drawn in Figure 6b. In fact, during LFR, when the laser focuses on a certain location, the powder there is heated and rapidly melted. As the laser moves away, the melted material at this location cools rapidly. The material at this location will undergo a complex cycle of repeated heating or even remelting as the next layer of cladding takes place. Any position within the LFRed coating, except for the final solidification areas on the top, will follow this process. The peak temperature of the thermal cycle changes continuously along with the distance from the already solidified position to the molten pool. When the distance is close (for example, the next layer), the reheating temperature of the previous position will be high, whereas when the molten pool is further away, the reheating temperature will be low. As for region A in Figure 6a, when the molten pool is in region A, thermal cycle 1 occurs, whereas when the molten pool is in region B, thermal cycle 2 occurs, and so on. These five cycles correspond to exactly five different distance ranges. This process of repeatedly rapid heating and cooling carries a high risk of cracking [34–37]. It is worth noting that the different regions shown in Figure 6a are distributed in three dimensions. In addition, the regions A, B, C, D, and E are schematic and should actually be much larger than those in Figure 6a.

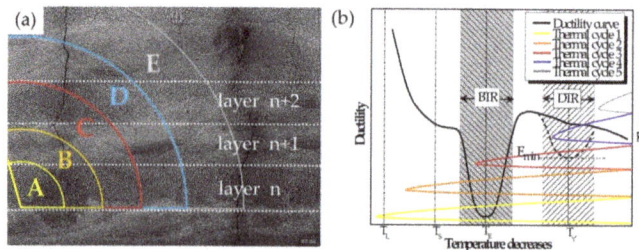

Figure 6. (**a**) Microstructure in the middle of the LFRed coating and (**b**) schematic diagram of the combined effect of ductility curve and thermal cycle. BTR—brittleness temperature region; DTR—ductility dip temperature region.

3.2.2. Cracking Mechanisms

In order to further explore the influence of composition and process on cracking sensitivity of LFRed K417G, a schematic diagram of the combined effect of ductility curve and the five thermal cycles is drawn. As shown in Figure 6b, the ductility of the material changes as the temperature decreases in the solidification process. There are two regions with obviously low ductility. One is the brittleness temperature region (BTR), which is prone to solidification cracking (SC) and liquation cracking (LC). The other is the ductility dip temperature region (DTR), which is prone to ductility dip cracking (DDC). The two low ductility regions are around the temperature of eutectic reaction temperature (T_E) and γ' precipitation temperature ($T_{\gamma'}$), respectively, whereas around other temperatures of liquidus temperature (T_L) and solidus temperature (T_S) show no low ductility regions. In addition, the material will undergo five types of thermal cycles, whose peak temperatures are above T_L, above T_S, above T_E, above $T_{\gamma'}$, and below $T_{\gamma'}$, respectively. The time of each thermal cycle is related to the process of LFR, while the ductility curve is mainly related to the composition of the superalloy. When the thermal cycle of the material is in two regions of BTR and DTR, the cracking behavior will be generated. The cracking behavior in K417G during LFR could be classified into three types. They are solidification cracking, liquation cracking, and ductility dip cracking.

Solidification Cracking

The schematic of the solidification cracking mechanism is shown in Figure 7. When the temperature drops below T_L, the primary solid phase of different orientations begins to form, as shown in Figure 7a. When the temperature drops below TS, these different orientations of solids grow alternately to form the grain skeleton. In this case, the residual liquid phase cannot flow freely between the solid dendrites, forming a closed and continuously distributed liquid film. Moreover, the rapid cooling rate results in the non-equilibrium solidification and enrichment of Al and Ti in the interdendritic zones, as shown in Figure 7b. When the content of (Al + Ti) in the liquid phase reaches the critical value, the eutectic reaction of L → γ + γ' will be generated between the dendrites, thus forming the eutectic structure. As shown in Figure 7c, the temperature at this time is TE. Under the action of contraction stress, continuous liquid film ends produce strain concentration, and it is easy to form shrinkage cavity and microcrack in the weak eutectic structure between the dendrites. In the process of continuous solidification (below the temperature of TE), if the interdendritic zone with shrinkage cavity and microcrack is at the grain boundary, the microcrack will propagate along the brittle grain boundary, thus forming the solidification crack. This cracking behavior occurs when the material experiences the thermal cycles of 1 and 2 (the yellow line and orange line in Figure 6b).

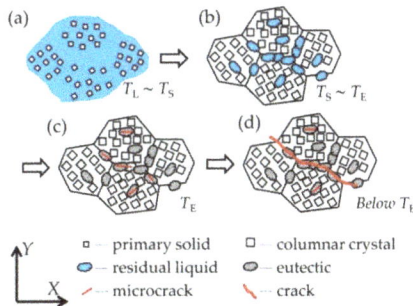

Figure 7. Schematic diagram of the formation process of solidification cracking.

In the process of crack growth, the residual stress is gradually released and the energy required for crack growth is gradually reduced. At the same time, as the deposited layer in the extension zone has solidified, the closer to the substrate, the more completely solidified, and the greater the strength of the intercrystalline bond. When the stress is not enough to break the intercrystalline bond, the crack will be terminated. In addition, as the number of deposited layers increases, the tensile stress in the repaired area gradually decreases and changes to compressive stress, and the crack expansion along the deposited direction towards the top will be impeded. Therefore, most solidification cracks will appear perpendicular to the direction of deposition. Figure 8a,b show images at a high magnification of a crack on the X–Y section. Figure 8c shows the morphology of the smooth area of the crack section, where there are almost no holes in the depth direction, which is determined by the nearly two-dimensional distribution of the interdendritic liquid film. Because the dendrites near the grain boundary remain free in the residual liquid phase and grow without solid obstacle, and the cracking is carried out along the grain boundary, the dendrites near the crack appear as round particles with a smooth surface, as shown in Figure 8d. Consequently, it can be seen that the crack in Figure 8 belongs to a solidification crack.

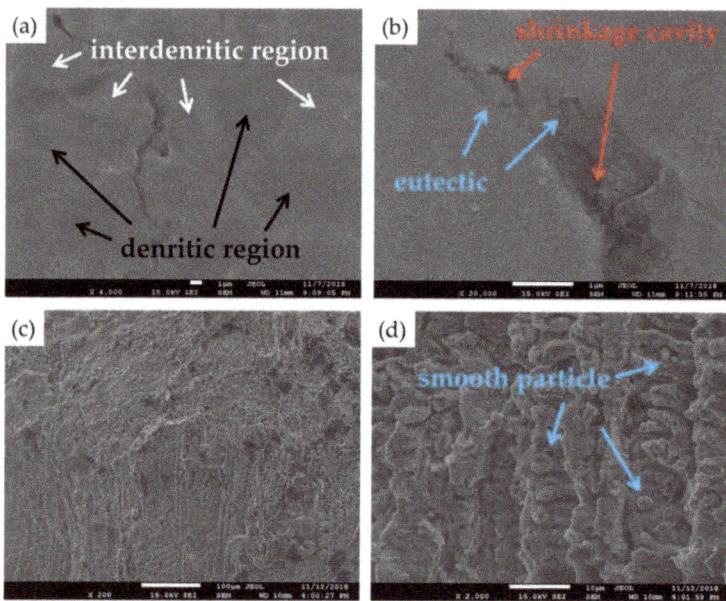

Figure 8. Morphology of (**a**,**b**) a solidification crack and (**c**,**d**) its section.

Liquation Cracking

Liquation cracking is the most common form of cracking behavior in superalloys with high (Al + Ti) content. Unlike solidification cracking, the source of liquation cracking is low-melting point eutectics in the heat affected zone (HAZ). These solidified eutectics will be re-melted to form liquid film when the reheated temperature is above TE, which will lead to intergranular cracking under the action of contraction force. This cracking process occurs during the thermal cycles of 3 (the red line Figure 6b). Figure 9 shows the formation process of liquation cracking. It is worth noting that the HAZ does not only exist in the substrate, but also in the previous layer when the laser is focusing on the current layer. In Figure 6a, for example, when the molten pool is in region C, the solidified region A will be a HAZ. Whereas when the molten pool is in region D, the solidified region B will be a HAZ at this time. In other words, there are countless HAZs inside the coating as long as the laser melting is not performed only once. To be exact, the HAZ shown in Figure 5 is the only original HAZ. With the continuous progress of LFR process, HAZs are constantly formed, and the eutectics with low-melting points in HAZs are constantly remelted, thus forming more and more crack sources. Because of the characteristic of epitaxial growth on the structure during the LFR process, the structure between layers has the genetic characteristic, which easily form columnar crystals with the same orientation. Consequently, the channels through the columnar dendrites between layers are generated. Once the liquefied microcrack is formed at the grain boundary, it will extend along the grain boundary with the accumulation of residual tensile stress and form the macroscopic crack through many deposited layers.

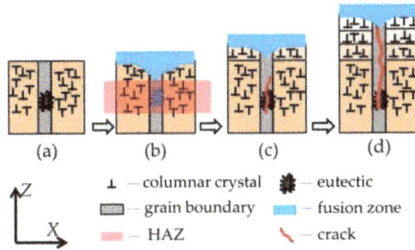

Figure 9. Schematic diagram of the formation process of liquation cracking, (**a**) a eutectic in a HAZ; (**b**) the liquefied eutectic; (**c**) the formation of a microcrack; (**d**) the extension of the microcrack.

Liquation cracks usually grow along the direction of dendrite growth, which is similar to an upward growth direction parallel to the deposition direction. As shown in Figure 10a, the extension direction of the crack seems to follow the growth law above. Moreover, it can be seen from Figure 10b that the crack is filled with broken liquid film, and eutectic and coarse γ' are distributed around the crack. Figure 10c shows the macroscopic morphology of the crack section. It has the characteristic of intergranular cracking. As shown in Figure 10d, obvious liquation of dendrite protrusions can be observed on the microscopic morphology of the crack section. The crack section takes on the shape of a potato of different sizes, indicating that it is the result of interdendritic liquid film separation and is a typical liquation crack. However, for the judgment of cracking mechanism of other cracks in Figure 5a, it is not sufficient to observe only the growth direction. Because of the complex solidification process of melting, remelting, partial remelting, cyclic annealing, and countless HAZ within the LFRed coating, it is difficult to identify whether the cracking behavior belongs to the liquation cracking or solidification cracking [38,39]. It can at least be determined that crack 1 in Figure 5a belongs to the liquation crack, as it exists in the original HAZ and is similar to an upward growth direction parallel to deposition.

Ductility Dip Cracking

When the material is experiencing thermal cycle 4 (the blue line Figure 6b), it will suffer the effect of continuously growing solid-phase shrinkage stress. No liquid phase exists in this process, and the deformation mode mainly depends on the vacancy diffusion or the dislocation climb along the grain boundary. When the processes of diffusion and climb are occurring, the dislocation will meet obstacles. If the ductility of the material is poor, cracking will occur due to the strain concentration. At this moment, the crack belongs to ductility dip crack. As these obstacles can be the vertex where the three grains intersect, or the carbides on the grain boundary, ductility dip cracking is generally generated by two types, as shown in Figure 11. The straight grain boundaries promote grain boundary sliding to form large voids and corresponding strain concentration at a vertex where the three grains intersect. These voids at the vertex eventually develop into a crack, which is type 1 in Figure 11. Type 2 in Figure 11 is mainly related to the carbides. In fact, the carbides here play a double role in the effect on the sensitivity of ductility dip cracking. On the one hand, the carbides lock the grain boundaries and reduce the grain boundary sliding, thus reducing the strain concentration at the vertex of three adjacent grains. The carbides, on the other hand, lock the grain boundaries but accumulate strain and voids around the carbides themselves, which may lead to the formation of cracks [40]. The effect of carbides on the sensitivity of ductility dip cracking depends on their type, size, and distribution.

The situation is very complicated. Emin in Figure 6b is the critical strain value of ductility dip cracking in the DTR, and its value can be used as an indicator to determine the sensitivity of ductility dip cracking of the material. When the carbides play a role in inhibiting ductility dip cracking, Emin will be slightly below the ductile curve, at which time the width of the DTR is very narrow or even does not exist.

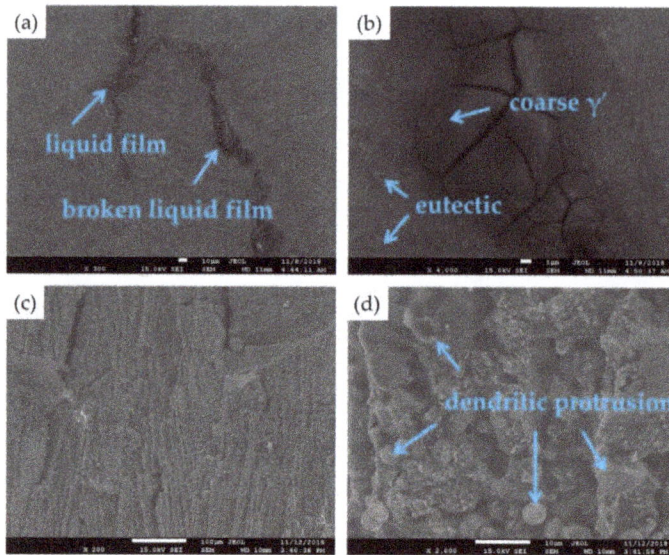

Figure 10. Morphology of (**a**,**b**) a liquation crack and (**c**,**d**) its section.

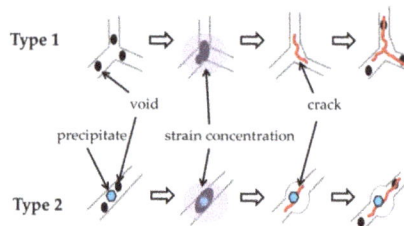

Figure 11. Schematic diagram of the formation process of ductility dip cracking.

The crack shown in Figure 12a passes through the vertex of the three adjacent grains and extends along the grain boundaries of these three grains. Moreover, the section in Figure 12b shows that no liquid film exists at the grain boundary of this crack. Consequently, it indicates that the crack should be a ductility dip crack formed in type 1 in Figure 11. However, because it is difficult to find ductility dip cracks formed in type 2, and if solidification cracking and grain boundary liquation cracking have occurred, the cracks expansion at this temperature will be further intensified or connected with ductility dip cracks. Therefore, the effect of the carbides on the sensitivity of ductility dip cracking in LFRed K417G is still unclear.

In summary, it can be determined that the cracking mechanism of cracks in LFRed K417G is solidification cracking, liquation cracking, and ductility dip cracking, respectively. However, the judgment of cracking mechanism of a crack requires comprehensive evidence, which should be combined with observation of macroscopic and microscopic morphology, determination of composition, and analysis of fracture, among others. Moreover, the judgment may still not be completely accurate, because sometimes it may be a combination of multiple cracking mechanisms. Fortunately, all cracking behaviors are related to the composition of the material and the process of LFR. Therefore, adjusting composition, optimizing process parameters, and post treatment can be utilized to control or eliminate cracking behavior.

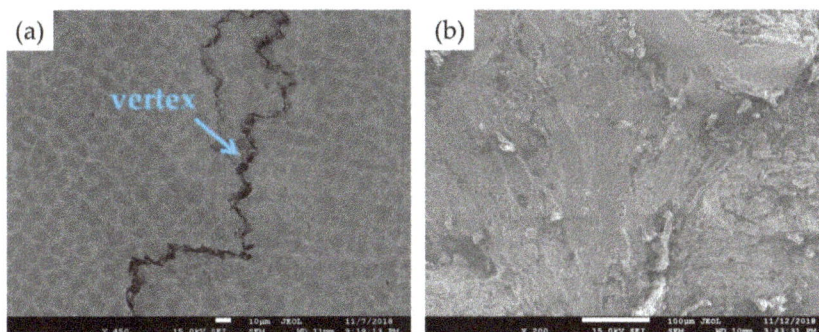

Figure 12. Morphology of (**a**) a ductility dip crack and (**b**) its section.

3.3. Effect of Laser Remelting Process on Microstructure and Cracking Behavior

3.3.1. Microstructure after Remelting

Laser remelting is considered as an effective post treatment to improve the quality and properties of LFRed coating. As shown in Figure 13, the surface quality of the coating after laser remelting is obviously superior to that before remelting. Laser remelting is also the process of reheating and solidification of the material. No new alloy powder is added during this process. Sharp points and small metal particles can absorb laser energy and remelt. It can be seen that a laser remelting process after laser deposition can make the surface smooth.

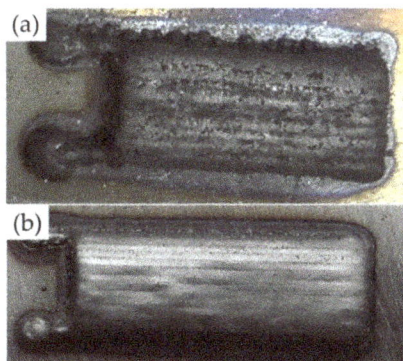

Figure 13. Surface morphology of the LFRed coating on K417G substrate (**a**) before and (**b**) after laser remelting.

Figure 14 shows the microstructure of LFRed K417G coating after laser remelting. The microstructure is still mainly composed of γ phase, γ' precipitated phase, $\gamma + \gamma'$ eutectic, and carbide. After remelting, the microstructure is obviously refined, and the size of the precipitate is reduced. It can be seen from the comparison between the line scan results in Figures 3 and 15 that the element segregation is somewhat reduced and the element distribution is more uniform.

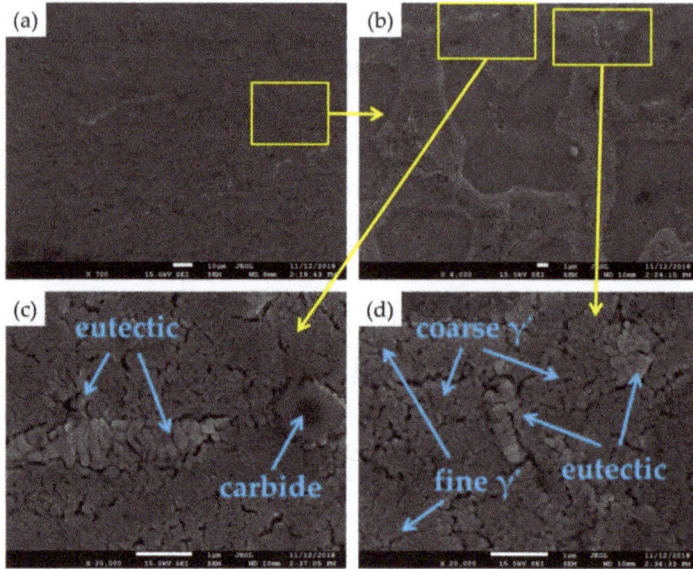

Figure 14. Typical microstructure on the X–Z section of the as-remelted LFRed K417G.

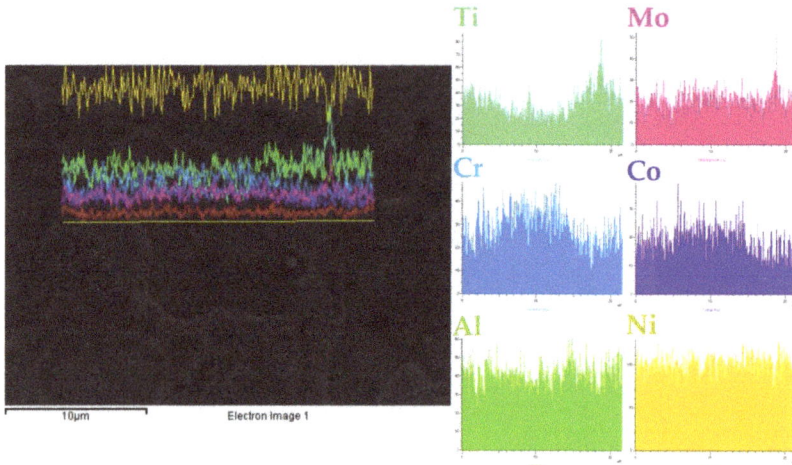

Figure 15. Energy dispersive spectrometer (EDS) results of line scanning in the local zone of Figure 14b.

3.3.2. Cracking Behavior after Remelting

Figure 16 shows the crack number and crack density of the three sections in the as-remelted K417G coating. According to the statistic of the crack count, the number of cracks in the coating after remelting does not decrease, but increases slightly, which is an undesirable phenomenon. However, the crack density decreases significantly, which is already less than 10^{-3} $\mu m/\mu m^2$. This indicates a significant reduction in the size of the cracks in the coating. Combined with the analysis of Section 3.2.2, it can be speculated that laser remelting plays a double role in cracking behavior. On the one hand, because of the effect of rapid heating and rapid solidification in the process of LFR, the material has experienced five types of thermal cycles (in Figure 6b), of which four can generate cracks. Laser remelting causes the material to undergo these thermal cycles again, thus doubling the susceptibility of cracking. Also, the thermal stress is higher because the heat is input again. On the other hand, laser remelting refines the microstructure, reduces elemental segregation, makes the precipitates more evenly distributed, and creates grains with a more different orientation, which undoubtedly hinder the connection of liquid film, thus hindering the extension of microcracks. For large cracks, especially the liquation cracks that have already formed in the HAZ, laser remelting is difficult to heal, and these cracks become larger as a result of higher thermal stress. As a result of this double action, the cracks continue to expand, the number of medium size cracks decreases, and the number of small size cracks increases.

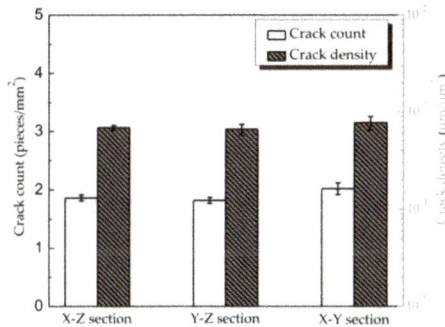

Figure 16. Crack count and density on three sections of as-remelted LFRed K417G.

3.4. Effect of Laser Remelting Process on Microhardness and Tribological Properties

3.4.1. Microhardness

Figure 17 shows the microhardness of the LFRed K417G coating before and after laser remelting process, from which it can be seen that the as-remelted coating has higher microhardness than that in the as-deposited coating, and the increase is about 50 HV0.2. Moreover, the microhardness of the as-deposited coating changes unsteadily, while the hardness distribution and the dispersion of each measuring point in the as-remelted coating are more uniform. According to the results discussed in Section 3.3, the microstructure after laser remelting is more refined; the distribution of precipitates is more uniform; the size and number of $\gamma + \gamma'$ eutectics, which are harmful to the improvement of hardness, are reduced; and more γ' are distributed in dendrites, which has a more significant enhancement effect. These combined effects result in hardness improvement and uniform distribution.

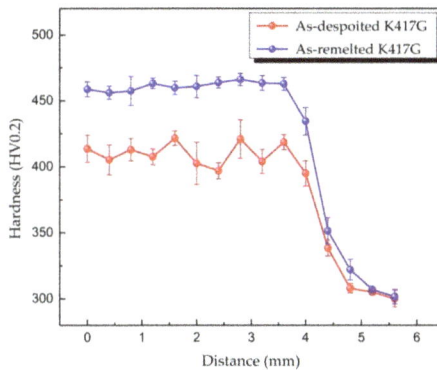

Figure 17. Microhardness of the LFRed K417G before and after laser remelting.

3.4.2. Tribological Properties

Friction coefficient evolutions of as-deposited K417G and as-remelted K417G are shown in Figure 18. The raw data have been smoothed (10 points Savitzky–Golay smooth) to facilitate the analysis of the results. At the beginning of the test, the friction coefficients of the two samples reach a high value. This indicates that the friction surface begins to show plastic deformation and adhesion since the beginning of the test, and under the effect of shear stress, the particles begin to peel off on the surface of the material. The average values of friction coefficients in the final 50 m sliding of the as-deposited and as-remelted samples are 0.68 and 0.54, respectively. It is worth noting that this friction coefficient is not a steady value. Because, at the end of the test, the two samples cannot reach the steady state, there are still large fluctuations. In general, the friction coefficient of the as-remelted sample is less than that of the as-deposited sample. Moreover, the friction coefficient of the as-remelted sample shows a downward trend in the late test period, while the as-deposited sample still fluctuates within a larger value.

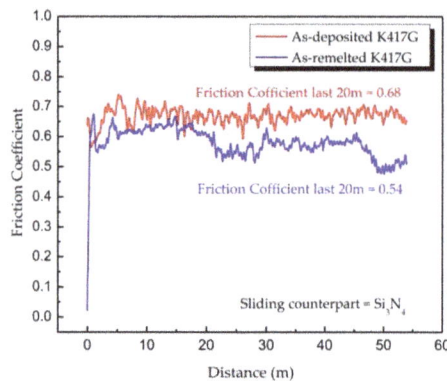

Figure 18. The friction coefficient of the LFRed coatings before and after laser remelting process.

As shown in Figure 19, two-dimensional profiles of the wear tracks are obtained at the end of tests. The raw data have been smoothed (10 points Savitzky–Golay smooth) as well. The result clearly shows that after 54 m of sliding, the depth and width of the wear tracks are much larger for the as-deposited K417G than for the as-remelted K417G. The wear rates of the as-deposited K417G and the as-remelted K417G are 5.95×10^{-14} and 3.24×10^{-14} mm^3/N·m, respectively. The wear rate of the coatings agrees well with the expected inverse proportional relation between this property and the hardness of the coatings. All these show that the as-remelted K417G has better tribological properties.

Figure 20 shows the wear surfaces of the as-deposited K417G and the as-remelted K417G. The two wear surfaces are both similar to the form of adhesion wear surface. Large areas of peeling and oxidation can be seen on the wear surfaces. As can be seen from Figure 20c,d, both wear surfaces have deep scratches. These scratches are not similar to those caused by abrasive particles. They are more likely to be caused by scraping of the abrasive chip that has flaked and adhered to the surface. In addition, it can be observed that the cracks on the surface of the as-remelted K417G are separated by the wear surface, while the cracks on the surface of the as-deposited K417G pass through the wear surface, showing that the cracks in the as-remelted K417G are shallower. This confirms, from another aspect, the effect of laser remelting on the reduction of crack size, as discussed in Section 3.3.2.

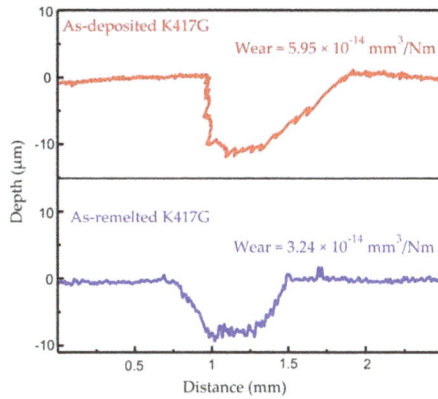

Figure 19. The 2D profiles of the wear track of the as-deposited K417G and as-remelted K417G tested against Si_3N_4 balls.

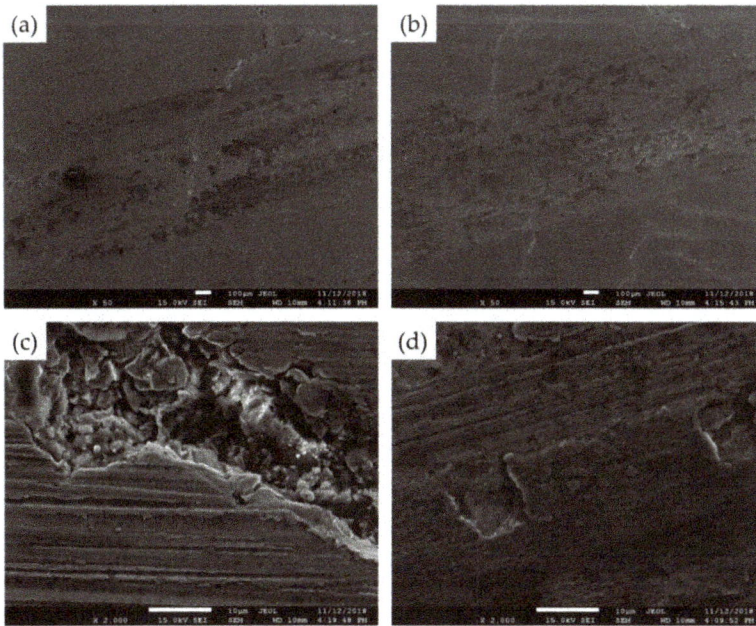

Figure 20. Wear surfaces of (**a,c**) the as-deposited K417G and (**b,d**) the as-remelted K417G.

4. Conclusions

The K417G Ni-based superalloy has been prepared on as-cast K417G substrate by the LFR process. The laser remelting process was applied as a post treatment to improve the properties of the LFRed coating. The main conclusions are as follows.

The microstructure of the LFRed K417G consists of γ phase, γ' precipitated phase, $\gamma + \gamma'$ eutectic, and carbide. The characteristic of cracking behavior is mainly influenced by the composition of K417G and the processing of LFR. Cracking mechanisms of the LFRed K417G include solidification cracking, liquation cracking, and ductility dip cracking.

Laser remelting can decrease the size of the cracks in the LFRed K417G, but not the number of cracks. After laser remelting, the microstructure of the coating was refined, and the element segregation was reduced. The as-remelted coating has higher microhardness, which can reach up to 460 HV0.2 compared with that of the as-deposited coating, and the increase is about 50 HV0.2. Similarly, the as-remelted K417G has a better tribological property than the as-deposited K417G. The wear surfaces are both related to adhesion wear.

Consequently, in the application of LFR technology to repair damaged K417G blades, laser remelting can be applied as an effective method for strengthening coating and as an auxiliary method for controlling cracking. However, cracks still exist. In order to eliminate cracking behavior, more efforts should be committed to component adjustment, process parameter optimization, and other post-treatment studies.

Author Contributions: Conceptualization, C.L.; Resources, J.L. and S.C.; Formal Analysis, Y.W. and H.Y.; Data Curation, X.Z. and S.L.; Validation, H.Y. and S.L.; Supervision, C.L.; Writing—Original Draft, S.L.

Funding: This research was funded by the Joint Founds of National Natural Science Foundation of China (NSFC)-Liaoning (No. U1508213) and National Key Research Project (No. 2017YFB0305801).

Acknowledgments: Bin Zhang and Jing Liang from Northeastern University are gratefully acknowledged for data analysis.

Conflicts of Interest: The authors declare no conflict of interest.

Abbreviations

LFR	Laser Forming Repairing
LFRed	Laser Forming Repaired
HAZ	Heat Affected Zone
RZ	Repaired Zone

References

1. Koh, K.H.; Griffin, J.H.; Filippi, S.; Akay, A. Characterization of turbine blade friction dampers. *J. Eng. Gas Turbines Power* **2004**, *127*, 856–862. [CrossRef]
2. Yang, C.X.; Xu, Y.L.; Nie, H.; Xiao, X.S.; Jia, G.Q.; Shen, Z. Effects of heat treatments on the microstructure and mechanical properties of Rene 80. *Mater. Des.* **2013**, *43*, 66–73. [CrossRef]
3. Yang, Y.H.; Xie, Y.J.; Wang, M.S.; Ye, W. Microstructure and tensile properties of nickel-based superalloy K417G bonded using transient liquid-phase infiltration. *Mater. Des.* **2013**, *51*, 141–147. [CrossRef]
4. Du, B.N.; Yang, J.X.; Cui, C.Y.; Sun, X.F. Effects of grain refinement on the microstructure and tensile behavior of K417G superalloy. *Mater. Sci. Eng. A* **2015**, *623*, 59–67. [CrossRef]
5. Pollock, T.M.; Tin, S. Nickel-based superalloys for advanced turbine engines: Chemistry, microstructure and properties. *J. Propul. Power* **2006**, *22*, 361–374. [CrossRef]
6. Gaumann, M.; Henry, S.; Cleton, F.; Wegniere, J.D.; Kurz, W. Epitaxial laser metal forming: Analysis of microstructure formation. *Mater. Sci. Eng. A* **1999**, *271*, 232–241. [CrossRef]
7. Lin, X.; Yue, T.M.; Yang, H.O.; Huang, W.D. Microstructure and phase evolution in laser rapid forming of a functionally graded Ti–Rene88DT Alloy. *Acta Mater.* **2006**, *54*, 1901–1915. [CrossRef]
8. Liu, F.C.; Lin, X.; Huang, C.P.; Song, M.H.; Yang, G.L.; Chen, J.; Huang, W.D. The effect of laser scanning path on microstructures and mechanical properties of laser solid formed Nickel-base superalloy Inconel 718. *J. Alloys Compd.* **2011**, *205*, 4505–4509. [CrossRef]
9. Ojo, O.A.; Richards, N.L.; Chaturvedi, M.C. Contribution of constitutional liquation of gamma prime precipitate to weld HAZ cracking of cast Inconel 738 superalloy. *Scr. Mater.* **2004**, *50*, 641–646. [CrossRef]
10. Ojo, O.A.; Chaturvedi, M.C. On the role of liquated γ' precipitates in weld heat affected zone microfissuring of a Nickel-based superalloy. *Mater. Sci. Eng. A* **2005**, *403*, 77–86. [CrossRef]
11. Li, Q.G.; Lin, X.; Wang, X.H.; Yang, H.O.; Song, M.H.; Huang, W.D. Research on the grain boundary liquation mechanism in heat affected zones of laser forming repaired K465 Nickel-based superalloy. *Metals* **2016**, *6*, 64. [CrossRef]

12. Tancret, F. Thermo-Calc and Dictra simulation of constitutional liquation of gamma prime (γ′) during welding of Ni base superalloys. *Comput. Mater. Sci.* **2007**, *41*, 13–19. [CrossRef]

13. Li, X.L.; Liu, J.W.; Zhong, M.L. Research on laser cladding superalloy K403. *Appl. Laser* **2002**, *22*, 283–286.

14. Zhou, Z.H.; Zhu, B.D. The study on the laser cladding process and cracking of cast Ni-based superalloy K3. *J. Mater. Eng.* **1996**, *1*, 32–35.

15. Yang, J.J.; Li, F.Z.; Wang, Z.M.; Zeng, X.Y. Cracking behavior and control of Rene 104 superalloy produced by direct laser fabrication. *J. Mater. Process. Technol.* **2015**, *225*, 229–239. [CrossRef]

16. Zhou, S.F.; Xu, Y.B.; Liao, B.Q.; Sun, Y.J.; Dai, X.Q.; Yang, J.X.; Li, Z.Y. Effect of laser remelting on microstructure and properties of WC reinforced Fe-based amorphous composite coatings by laser cladding. *Opt. Laser Technol.* **2018**, *103*, 8–16. [CrossRef]

17. Gao, W.Y.; Zhao, S.S.; Wang, Y.B.; Liu, F.L.; Zhou, C.Y.; Lin, X.C. Effect of re-melting on the cladding coating of Fe-based composite powder. *Mater. Des.* **2014**, *64*, 490–496. [CrossRef]

18. Zhang, Y.Y.; Lin, X.; Wang, L.L.; Wei, L.; Liu, F.G.; Huang, W.D. Microstructural analysis of Zr55Cu30Al10Ni5 bulk metallic glasses by laser surface remelting and laser solid forming. *Intermetallics* **2015**, *66*, 22–30. [CrossRef]

19. Wang, Q.Y.; Xi, Y.C.; Zhao, Y.H.; Liu, S.; Bai, S.L.; Liu, Z.D. Effects of laser re-melting and annealing on microstructure, mechanical property and corrosion resistance of Fe-based amorphous/crystalline composite coating. *Mater. Charact.* **2017**, *127*, 239–247. [CrossRef]

20. Li, R.F.; Jin, Y.J.; Li, Z.G.; Zhu, Y.Y.; Wu, M.F. Effect of the remelting scanning speed on the amorphous forming ability of Ni-based alloy using laser cladding plus a laser remelting process. *Surf. Coat. Technol.* **2014**, *259*, 725–731. [CrossRef]

21. AlMangour, B.; Grzesiak, D.; Yang, J. Scanning strategies for texture and anisotropy tailoring during selective laser melting of TiC/316L stainless steel nanocomposites. *J. Alloy. Compd.* **2017**, *728*, 424–435. [CrossRef]

22. AlMangour, B.; Grzesiak, D.; Cheng, J.; Ertas, Y. Thermal behavior of the molten pool, microstructural evolution, and tribological performance during selective laser melting of TiC/316L stainless steel nanocomposites: Experimental and simulation methods. *J. Mater. Process. Technol.* **2018**, *257*, 288–301. [CrossRef]

23. Tabernero, I.; Lamikiz, A.; Martinez, S.; Ukar, E.; Figueras, J. Evaluation of the mechanical properties of Inconel 718 components built by laser cladding. *Int. J. Mach. Tool. Manuf.* **2011**, *51*, 465–470. [CrossRef]

24. Choi, J.P.; Shin, G.H.; Yang, S.G.; Yang, D.Y.; Lee, J.S.; Brochu, M.; Yu, J.H. Densification and microstructural investigation of Inconel 718 parts fabricated by a selective laser melting. *Powder Technol.* **2017**, *310*, 60–66. [CrossRef]

25. Li, Q.G.; Lin, X.; Liu, F.C.; Huang, W.D. Microstructural characteristics and mechanical properties of laser solid formed K465 superalloy. *Mater. Sci. Eng. A* **2017**, *700*, 649–655. [CrossRef]

26. Montazeri, M.; Ghaini, F.M. The liquation cracking behavior of IN738LC superalloy during low power Nd:YAG pulsed laser welding. *Mater. Charact.* **2012**, *67*, 65–73. [CrossRef]

27. Gong, L.; Chen, B.; Du, Z.H.; Zhang, M.S.; Liu, R.C.; Liu, K. Investigation of solidification and segregation characteristics of cast Ni-Base superalloy K417G. *J. Mater. Sci. Technol.* **2018**, *34*, 541–550. [CrossRef]

28. Gong, L.; Chen, B.; Yang, Y.Q.; Du, Z.H.; Liu, K. Effect of N content on microsegregation, microstructure and mechanical property of cast Ni-base superalloy K417G. *Mater. Sci. Eng. A* **2017**, *701*, 111–119. [CrossRef]

29. Basak, A.; Das, S. Microstructure of nickel-base superalloy MAR-M247 additively manufactured through scanning laser epitaxy (SLE). *J. Alloy. Compd.* **2017**, *705*, 806–816. [CrossRef]

30. Carter, L.N.; Martin, C.; Withers, P.J.; Attallah, M.M. The influence of the laser scan strategy on grain structure and cracking behavior in SLM powder-bed fabricated nickel superalloy. *J. Alloy. Compd.* **2014**, *615*, 338–347. [CrossRef]

31. Ghosh, S.K.; Partha, S. Crack and wear behavior of SiC particulate reinforced aluminium based metal matrix composite fabricated by direct metal laser sintering process. *Mater. Des.* **2011**, *32*, 139–145. [CrossRef]

32. Zhao, X.M.; Chen, J.; Lin, X.; Huang, W.D. Study on microstructure and mechanical properties of laser rapid forming Inconel 718. *Mater. Sci. Eng. A* **2008**, *478*, 119–124. [CrossRef]

33. Ola, O.T.; Ojo, O.A.; Chaturvedi, M.C. Role of filler alloy composition on laser arc hybrid weldability of nickel-base IN738 superalloy. *Mater. Sci. Technol.* **2014**, *30*, 1461–1469. [CrossRef]

34. Dadbakhsh, S.; Hao, L. Effect of hot isostatic pressing (HIP) on Al composite parts made from laser consolidated Al/Fe₂O₃ powder mixtures. *J. Mater. Process. Technol.* **2012**, *212*, 2474–2483. [CrossRef]

35. Wei, K.W.; Gao, M.; Wang, Z.M.; Zeng, X.Y. Effect of energy input on formability, microstructure and mechanical properties of selective laser melted AZ91D magnesium alloy. *Mater. Sci. Eng. A* **2014**, *611*, 212–222. [CrossRef]

36. Alimardani, M.; Fallah, V.; Iravani-Tabrizipour, M.; Khajepour, A. Surface finish in laser solid freeform fabrication of an AISI 303L stainless steel thin wall. *J. Mater. Process. Technol.* **2012**, *212*, 113–119. [CrossRef]

37. Hussein, A.; Hao, L.; Yan, C.; Everson, R. Finite element simulation of the temperature and stress fields in single layers built without-support in selective laser melting. *Mater. Des.* **2013**, *52*, 638–647. [CrossRef]

38. Zhou, Z.P.; Huang, L.; Shang, Y.J.; Li, Y.P.; Jiang, L.; Lei, Q. Causes analysis on cracks in nickel-based single crystal superalloy fabricated by laser powder deposition additive manufacturing. *Mater. Des.* **2018**, *160*, 1238–1249. [CrossRef]

39. Ojo, O.A.; Richards, N.L.; Chaturvedi, M.C. Microstructural study of weld fusion zone of TIG welded IN 738LC nickel-based superalloy. *Scr. Mater.* **2004**, *51*, 683–688. [CrossRef]

40. Ramirez, A.J.; Lippold, J.C. High temperature behavior of Ni-base weld metal Part II—Insight into the mechanism for ductility dip cracking. *Mater. Sci. Eng. A* **2004**, *380*, 245–258. [CrossRef]

coatings

MDPI

Article

Towards Functional Silicon Nitride Coatings for Joint Replacements

Luimar Filho [1], Susann Schmidt [2,3], Klaus Leifer [1], Håkan Engqvist [1], Hans Högberg [3] and Cecilia Persson [1,*]

[1]   Division of Applied Materials Science, Department of Engineering Sciences, Uppsala University, Uppsala 752 37, Sweden; luimar.filho@angstrom.uu.se (L.F.); klaus.leifer@angstrom.uu.se (K.L.); hakan.engqvist@angstrom.uu.se (H.E.)

[2]   IHI Ionbond AG, Industriestrasse 211, Olten 4600, Switzerland; susann.schmidt@ionbond.com

[3]   Thin Film Physics Division, Department of Physics, Chemistry and Biology (IFM), Linköping University, Linköping 581 83, Sweden; hans.hogberg@liu.se

*   Correspondence: cecilia.persson@angstrom.uu.se; Tel.: +46-184-717-911

Received: 30 November 2018; Accepted: 18 January 2019; Published: 25 January 2019

Abstract: Silicon nitride (SiN_x) coatings are currently under investigation as bearing surfaces for joint implants, due to their low wear rate and the good biocompatibility of both coatings and their potential wear debris. The aim of this study was to move further towards functional SiN_x coatings by evaluating coatings deposited onto CoCrMo surfaces with a CrN interlayer, using different bias voltages and substrate rotations. Reactive direct current magnetron sputtering was used to coat CoCrMo discs with a CrN interlayer, followed by a SiN_x top layer, which was deposited by reactive high-power impulse magnetron sputtering. The interlayer was deposited using negative bias voltages ranging between 100 and 900 V, and 1-fold or 3-fold substrate rotation. Scanning electron microscopy showed a dependence of coating morphology on substrate rotation. The N/Si ratio ranged from 1.10 to 1.25, as evaluated by X-ray photoelectron spectroscopy. Vertical scanning interferometry revealed that the coated, unpolished samples had a low average surface roughness between 16 and 33 nm. Rockwell indentations showed improved coating adhesion when a low bias voltage of 100 V was used to deposit the CrN interlayer. Wear tests performed in a reciprocating manner against Si_3N_4 balls showed specific wear rates lower than, or similar to that of CoCrMo. The study suggests that low negative bias voltages may contribute to a better performance of SiN_x coatings in terms of adhesion. The low wear rates found in the current study support further development of silicon nitride-based coatings towards clinical application.

Keywords: silicon nitride; coating; reactive high-power impulse magnetron sputtering; wear; joint replacements

---

1. Introduction

The need for improved materials for biomedical applications is continuously growing as a result of the increasingly active, ageing population [1–3]. Total joint replacements, such as total hip or knee replacements (THR and TKR, respectively) are commonly used in arthritic, pain-ridden patients [4], with success rates of over 90% after 10 years [1,2,5]. THRs typically consist of a ball-and-cup configuration, featuring a CoCrMo or ceramic [1] ball sliding against a polyethylene cup. Ultra-high molecular weight polyethylene (UHMWPE) has been a widely used cup material, and the recent introduction of cross-linked polyethylene (XLPE) appears to permit even lower polymer wear [6–8]. This is of great importance, since wear debris can result in inflammation, osteolysis, and loosening of the prosthesis [1,9]. Furthermore, metallic ion release and debris [10] can cause metallosis and the formation of pseudo-tumors [11]. One possibility for reducing the metallic ion release and wear is to apply a ceramic coating on the bearing surface [12–14]. For example, TiN and ZrN coatings are currently available on the market for knee replacements, particularly aimed at hypersensitive patients [15,16]. Another possible use of ceramic coatings in joint implants is at the taper junction, which is notoriously prone to corrosion [17,18].

In this work, we investigate silicon nitride (SiN$_x$) coatings for joint implants. This type of coating has previously shown promising properties in terms of high biocompatibility, hardness, and low wear rates [17–20]. A further possible advantage of this coating compared to other ceramic coatings is its slow solubility in aqueous solutions [21,22], in combination with the high biocompatibility of its wear particles and ions [23]. We seek to develop SiN$_x$ coatings that dissolve controllably and generate wear particles of a higher dissolution rate than the coating itself (due to the higher surface area), and whose dissolution also gives biocompatible ions. This, in turn, may give rise to a less negative biological response compared to other materials' wear particles. Prior work on these coatings has been focused on depositing coatings without rotation on flat, 2D substrates. As sputtering is a line-of-sight-deposition technique, the coating of 3D details such as implants requires rotation during processing. The choice of parameters for rotating the substrate table during deposition (i.e., static, 1-fold, or 3-fold substrate rotation) influences the direction (angle), the flux of sputtered material, and the energy distribution of the sputtered material arriving to the substrate [24,25]. It is necessary to control these parameters, as the properties of the sputtered material will ultimately determine the nucleation and adhesion to the substrate, as well as the initial and continued growth of the film. To secure a good adhesion between the metallic 3D substrate and the ceramic SiN$_x$ coating, we applied growth by reactive high-power impulse magnetron sputtering (rHiPIMS), as this technique is favorable in forming a high energy flux of the sputtered material [26,27]. In addition, and for this demanding application, we applied an interlayer to promote chemical bonding in the interface region between the substrate and the coating [28].

The aim of this study was to move towards functional SiN$_x$ coatings on 3D implants by evaluating newly developed coatings deposited onto CoCrMo with a CrN interlayer, using different bias voltages, and 1- or 3-fold rotation of the substrate. The surface roughness, coating adhesion, and chemical composition throughout the coating layers and substrate were evaluated. The coefficient of friction and wear rate against bulk Si$_3$N$_4$ were evaluated in reciprocal ball-on-disc tests.

2. Materials and Methods

2.1. Coating Preparation

Direct current magnetron sputtering processes were used to coat mirror-polished CoCrMo discs ([29], R_a < 12 nm) with a CrN interlayer, followed by a SiN$_x$ top layer, deposited using rHiPIMS (CC800/9 ML(CemeCon AG Würselen, Germany)) in a mixed N$_2$ and Ar atmosphere. Substrates were mounted in a 1-fold (1f) or 3-fold (3f) rotational set-up. The specific process settings are shown in Table 1. The interlayer was deposited using a pressure of 200 MPa at negative bias voltages of 100 V (Low), 300 V (Medium), and 900 V (High). The SiN$_x$ coatings were deposited at a pressure of 600 MPa, with an average discharge power of 3300 W using a pulse frequency of 800 Hz, and a pulse width of 200 μs in a N$_2$/Ar mixture with a flow ratio of 0.27.

2.2. Compositional Analysis

The chemical composition of the coatings was analyzed by X-ray photoelectron spectroscopy (XPS) (Axis UltraDLD, Kratos Analytical, Manchester, UK) using monochromatic Al (Kα) X-ray radiation (hv = 1486.6 eV). The pressure in the analysis chamber during acquisition was less than 1×10^{-7} Pa. Samples were sputter cleaned for 120 s with a 2 keV Ar$^+$ ion beam. The Ar$^+$ beam was rastered over an area of 3×3 mm^2 at an incidence angle of 20°. Sputter cleaning was carried out to remove the surface oxygen layer and carbon due to air exposure. Automatic charge compensation was applied throughout the acquisition, owing to the electrical insulating nature of the SiN$_x$ coatings. The core level spectra recorded after Ar^+ sputter cleaning was used to determine the chemical composition of the SiN_x coatings, using a Shirley-type background together with elemental cross sections provided by Kratos Analytical.

2.3. Cross-Section Characterization

The coating microstructure, thickness, and composition were evaluated throughout the coating layers. A cross-section of the discs was initially sputtered by sputter coater (Au/Pd) for 30 s to reduce the charging effect, before being prepared using a focused ion beam (FIB; FEI Strata DB235, FEI, Hillsboro, OR, USA), with a platinum layer of 1 μm deposited on top to minimize the damage caused by the ion beam. The milling steps were from 7000 to 500 pA at 30 kV. Additional chemical analysis of the interlayers was undertaken using a scanning electron microscope (Zeiss Merlin with AZtec EDS/EBSD, Oberkochen, Germany) at 20 kV equipped with EDS Silicon Drift Detector AZtec (INCA energy) software (AZTEC 3.3 SP1, Oxford Instruments, High Wycombe, UK).

2.4. Surface Roughness

To evaluate the surface roughness, Vertical Scanning Interferometry (VSI), was used (WYKO NT1110, Vecco, Tucson, AZ, USA). The analyzed area was 736×480 μm^2 with an objective lens of $10\times$ and Field of View (FOV) of $0.5\times$. Three measurements were performed on each sample. From these, the arithmetic average of the absolute values (R_a) was obtained.

2.5. Adhesion

The coating adhesion was evaluated by Rockwell indentation testing, using an applied load of 100 N with the Rockwell tip (CA1819) in three different locations according to ISO 26443-2008 [30].

2.6. Wear Resistance

Wear tests of the coatings were performed against 10 and 20 mm diameter bulk Si_3N_4 balls, manufactured according to [31]. Si_3N_4 was chosen as the counter surface to provide a worst-case scenario, hard-on-hard contact, and to simulate the coating run against itself. This facilitates a comparison with earlier work, where such a contact was used [17,19,32]. CoCrMo alloy discs, manufactured according to [33], with 21.9 mm diameter and 5 mm thickness were used as controls. The tests were performed in an in-house reciprocating ball-on-disc wear test machine, as shown in Figure 1. The applied loads were 1, 2.44, and 2.45 N, giving a maximum Hertzian contact pressure of approximately 328 MPa (20 mm ball), 442 MPa (20 mm ball), and 700 MPa (10 mm ball). The typical contact pressure in a ceramic-on-ceramic (COC) prosthesis has been estimated to 90 MPa [34]. However, in the case of edge loading by micro-separation in a ceramic-on-metal (COM) prosthesis, a contact pressure of approximately 700 MPa has been estimated, which could be considered a worst case scenario [35]. A frequency of 1 Hz and a stroke length of 10 mm was used to produce three parallel wear tracks on each sample, running for 10,000 cycles. Specimens were kept in a heated Polytetrafluoroethylene (PTFE) container with a bath at temperature of 37 ± 3 °C throughout the test. To simulate body fluid, 25 vol.% fetal bovine serum was used (FBS, GE Healthcare Hyclone, EU approved, origin South America, Chicago, IL, USA), complemented with 0.075 wt.% sodium azide (Sigma-Aldrich, St. Louis, MO, USA, S8032-25G) and 20.0 mM ethylene–diaminetetraacetic acid solution (EDTA, Sigma-Aldrich, 03690), according to [36].

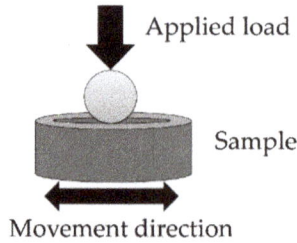

Figure 1. Schematic of the reciprocating ball-on-disc set-up for wear testing.

The specific wear rate was calculated following Archard's wear equation [37]:

$$\text{Specific wear rate} = \frac{\text{Wear volume}}{\text{Load} \times \text{sliding distance}} \tag{1}$$

where the wear volume was estimated from the cross-sectional area at the initial, middle, and final parts of the wear track, as measured with VSI after the reciprocal wear test.

2.7. Statistical Analysis

IBM SPSS Statistics v 22 was used for all statistical analyses. Welch's robust test for analysis of variance was performed (Levene's test for homogeneity of variances was significant for most analyses), followed by a Dunnett T3 post-hoc test. A critical level of $\alpha = 0.05$ was used to determine significance.

3. Results

3.1. Microstructure, Coating Thickness, and Composition

SEM cross-sections of coatings deposited using 1-fold rotation showed dense coatings (cf. Figure 2a–c), while the density, as judged upon appearance, decreased as 3-fold rotation was applied (cf. Figure 2d). The thicknesses and chemical composition of the coatings and interlayers are summarized in Table 1. A SiN_x thickness between 4.2 and 4.4 μm was measured. The coatings showed a Si content of between 43 at.% and 46 at.%, while N contents were between 50 at.% and 53 at.% and O contents of 2.1 at.%–2.2 at.% were measured, accounting for comparatively high N/Si ratios of 1.10–1.25.

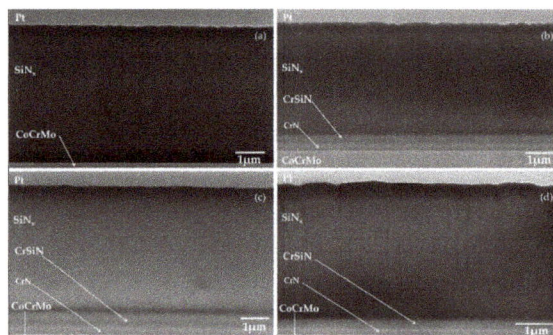

Figure 2. Cross-sections of (**a**) CoCr-SiN_x, without CrN interlayer, a dense coating could be observed; (**b**) CoCr-CrN(H)-SiN_x, deposited using 1-fold rotation and a negative bias voltage of 900 V; (**c**) CoCr-CrN(L)-SiN_x, with 1-fold rotation and a negative bias of 100 V; and (**d**) CoCr-CrN(L)-SiN_x-3f, deposited using 3-fold rotation and a negative bias of 100 V.

Table 1. The investigated materials, coating thicknesses, and chemical composition of the SiN_x top layers as measured by XPS.

Material	Interlayer	SiN_x Thickness (nm)	Interlayer Thickness (nm)	Top Layer Composition			
				Si (at.%)	N (at.%)	O (at.%)	N/Si Ratio
CoCr	–	–	–	–	–	–	–
CoCr-SiN$_x$	–	4360	–	46	50	2.1	1.10
CoCr-CrN(L)-SiN$_x$	CrN	4260	810	44	52	2.2	1.11
CoCr-CrN(M)-SiN$_x$	CrN	4400	750	44	53	2.1	1.19
CoCr-CrN(H)-SiN$_x$	CrN	4400	650	43	53	2.2	1.25
CoCr-CrN(L)-SiN$_x$-3f	CrN	4200	500	45	51	–	1.11

3.2. Surface Roughness

According to the standard for biomedical implants [38], the average surface roughness (R_a) of bearing surfaces or femoral heads of metallic or ceramic surfaces must be below 50 nm. Both coated and uncoated samples exhibited a surface roughness below this requirement (16–33 nm and 7 nm, respectively). As reported in Table 2, the roughness of the uncoated CoCrMo surface was significantly lower than that of the coated surfaces ($\alpha < 0.05$). The coating without an interlayer was significantly smoother (R_a = 16.3 nm) than the coatings deposited with an interlayer bias voltage (100, 300, and 900 V, $\alpha < 0.02$, R_a ranging between 23.9 and 32.8 nm). No significant difference in R_a was observed for coatings deposited at the low bias voltage (L) in either 1f or 3f rotations ($\alpha > 0.05$, R_a = 25.2 and 23.9 nm on average, respectively).

Table 2. Surface roughness and adhesion of the investigated materials.

Material	R_a (nm)	HRC Adhesion, (ISO Class)
CoCr	6.6 ± 0.4	N/A
CoCr-SiN$_x$	16.3 ± 1.8	3
CoCr-CrN(L)-SiN$_x$	25.2 ± 0.1	1
CoCr-CrN(M)-SiN$_x$	26.1 ± 1.2	2
CoCr-CrN(H)-SiN$_x$	32.8 ± 2.7	2
CoCr-CrN(L)-SiN$_x$-3f	23.9 ± 0.7	0–1

3.3. Adhesion

Results for the Rockwell C (HRC) adhesion tests are presented in Table 2. Improved values for coatings featuring a CrN interlayer were found compared to those without. Furthermore, a lower bias voltage during the CrN interlayer formation appeared to benefit the adhesion. For coatings deposited under 3-fold rotation, the adhesion improved further, to an HRC ISO class 0–1.

3.4. Wear Resistance

The initial stage of the experiment showed a higher friction coefficient (0.78 and 0.48 at 700 MPa, 0.51 and 0.39 at 442 MPa, and 0.48 and 0.33 at 328 MPa) for coated and non-coated samples respectively, which stabilized after 2000 cycles. The coefficient of friction was therefore averaged after the running-in phase, i.e., for the cycles between 2000 and 10,000, and is reported in Figure 3. The friction coefficients against the coatings were slightly higher than those against CoCr. However, a statistically significant difference was only found between CoCr and CoCr-CrN(H)-SiN$_x$ (α = 0.04 at the 328 MPa contact pressure).

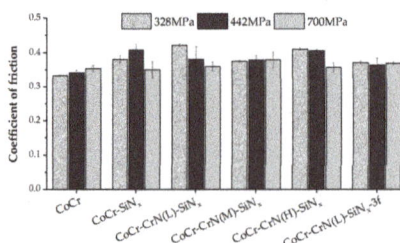

Figure 3. Coefficient of friction at estimated contact pressures of 328 and 442 MPa, using a ball diameter of 20 mm, with applied loads of 1 and 2.45 N, respectively. The coefficient of friction at an estimated contact pressure of 700 MPa is also included, using a ball diameter of 10 mm and an applied load of 2.44 N.

For the ball of diameter 20 mm, CoCr showed a higher specific wear rate compared to coated samples at the lower contact pressure of 328 MPa, but a similar wear rate at the higher contact pressure of 442 MPa. For the 442 MPa contact pressure, a significant difference could be found only between CoCr-SiN$_x$ and CoCr-CrN(L)-SiN$_x$-3f (α = 0.015), with the latter giving the lowest specific wear rate of all samples. For the smaller ball diameter (10 mm), CoCr again showed a higher wear rate than the coated samples, as shown in Figure 4. All wear tracks were analyzed by VSI and reported on Figure A1.

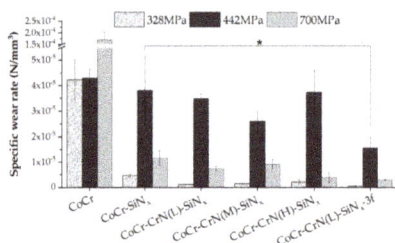

Figure 4. Specific wear rates at 328 and 442 MPa, ran against a 20 mm Si$_3$N$_4$ ball, as well as at 700 MPa with 10 mm ball. * Statistically significant difference.

4. Discussion

This study evaluated the effects of: (i) The addition of a CrN interlayer prior to depositing the SiN$_x$ coating; (ii) different bias voltages during interlayer formation; and (iii) the level of rotation during deposition on coating properties and wear performance.

Coatings deposited by 1f rotation presented uniform, featureless, and dense structures, while coatings deposited in a 3-fold rotation set-up showed a comparatively facetted growth with visible columns, as illustrated in Figure 2. This is related to the arrival rate of film-forming species as well as their energy during coating growth, where samples mounted in the 1f rotational set-up are exposed to more and higher energetic film forming species per time unit [39–41].

The coatings showed a high N content, with N/Si ratios between 1.10 and 1.25. In earlier studies, this has been found to be beneficial, in terms of a lower dissolution rate [21] and improved mechanical properties [42].

The average surface roughness of the coatings as deposited ranged between 16 and 23 nm, fulfilling the standard requirements for joint bearing surfaces, although post-polishing would likely be employed for commercial purposes. These higher roughness values as compared to the non-coated surface indicate the existence of coating facets, as shown in Figure 2. Other coatings for joint implants deposited using rHiPIMS (but using different deposition parameters) have shown similar roughness values [26,28].

The coating adhesion was improved by the implementation of a CrN interlayer, and was better for the 3-fold coatings compared to the 1-f coatings, with the latter result likely arising from a lower coating density giving lower residual stresses, as a result of the increased rotation [25]. The enhancement in adhesion after CrN addition was comparable to previous work [43,44].

The coefficient of friction of 0.33–0.42 is in the range of those reported in earlier, comparable wear studies on ceramic-on-ceramic implants (0.25–0.8) [45], but in the higher range of our earlier studies on SiN_x coatings [17,32]. The specific wear rates were also higher in general, except for CoCr at 328 MPa and 442 MPa [17,32]. These differences may be due to differences in coating morphology and density, counter surface ball sizes, and, in some cases, higher loads applied in the current study [46]. Other experimental coatings for joint implants have shown specific wear rates in the same order of magnitude under similar tests conditions. However direct comparisons are difficult due to variations in the set-up [17,32,47–49]. Formation of a tribofilm between the surfaces may occur during wear tests of these materials, resulting in a lubrication effect that reduces the wear and the coefficient of friction [50–52]. However, this was not investigated herein.

Wear rates at the lowest load show promising results for both ball sizes, while coating wear rates approached those of CoCr for the higher load (larger ball size). The highest contact pressure was chosen based on simulated micro separation studies for COC contacts [53]. Even higher contact pressures of around 1 GPa have, however, been reported in other studies for edge loading conditions [35]. Therefore, for hard-on-hard contacts, further testing should be considered at higher contact pressures. However, it should be noted that these cases are to be considered as worst-case scenarios. In fact, most joint prostheses currently implanted use a hard-on-soft bearing, i.e., a metal or ceramic actuating against a polymer. Therefore, further testing against polymeric surfaces would be of high interest. In addition, tests performed in a setting more similar to that of the end application, such as in joint simulators, and for longer periods of time, are needed.

5. Conclusions

SiN_x coatings were deposited onto CoCrMo substrates by rHiPIMS, under 1- and 3-fold rotation, to evaluate the resulting properties for possible application in joint implants. It could be concluded that:

- 3-fold deposition gave rise to less dense coatings as compared to 1-fold rotation, as demonstrated by the FIB cross-sections;
- The deposition method resulted in wear-resistant coatings, with no wearing down or flaking off when run against Si_3N_4, likely due to their high N content as well as their relatively high density;
- Despite the relatively higher roughness and lower density, the CoCr-CrN(L)-SiN_x-3f coating, deposited at a lower bias voltage and by 3-fold rotation, presented the lowest specific wear rate against Si_3N_4 balls, also compared to a CoCrMo control.

The present results indicate some promising properties of these coatings, although further studies are needed, especially in a 3D setting on full hip implants.

Author Contributions: Conceptualization, S.S., H.E., H.H. and C.P.; Methodology, L.F., S.S. and C.P.; Validation, L.F. and S.S.; Formal Analysis, L.F., S.S. and C.P.; Investigation, L.F. and S.S.; Resources, C.P., H.H., K.L. and H.E.; Writing—Original Draft Preparation, L.F., S.S. and C.P.; Writing—Review and Editing, L.F., S.S., K.L., H.E., H.H. and C.P.; Visualization, L.F., S.S. and C.P.; Supervision, K.L., H.E., H.H. and C.P.; Project Administration, H.H., H.E. and C.P.; Funding Acquisition, C.P., H.E., H.H. and K.L.

Funding: This research was funded by the European Union, No. FP7-NMP-2012-310477 (Life Long Joints project); the Erasmus Mundus Programme, Euro-Brazilian Windows + Project, No. 2014-0982; the Swedish Government Strategic Research Area in Materials Science on Advanced Functional Materials, No. 2009-00971.

Acknowledgments: Susan Peacock is gratefully acknowledged for proof-reading.

Conflicts of Interest: Håkan Engqvist is co-inventor on a patent related to similar coatings.

Appendix A

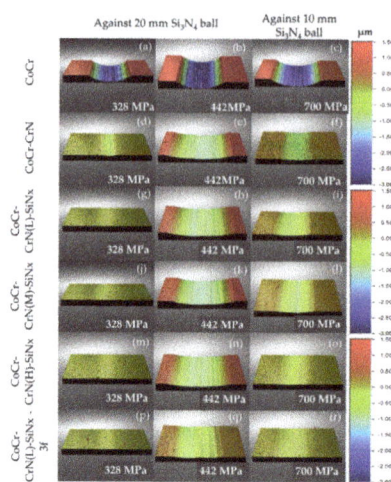

Figure A1. Typical wear tracks from reciprocal wear tests against Si$_3$N$_4$, as analyzed by Vertical Scanning Interferometry (VSI).

References

1. Kärrholm, J.; Lindahl, H.; Malchau, H.; Mohaddes, M.; Rogmark, C.; Rolfson, O. *Swedish Hip Arthroplasty Register Annual Report 2015*; Swedish Hip Arthroplasty Register: Gothenburg, Sweden, 2015.
2. Sundberg, M.; Lidgren, L.; W-Dahl, A.; Robertsson, O. *Swedish Knee Arthroplasty Register Annual Report 2015*; Swedish Hip Arthroplasty Register: Lund, Sweden, 2015.
3. Verheyen, C.C.; Verhaar, J.A. Failure rates of stemmed metal-on-metal hip replacements. *Lancet* **2012**, *380*, 105. [CrossRef]
4. Learmonth, I.D.; Young, C.; Rorabeck, C. The operation of the century: Total hip replacement. *Lancet* **2007**, *370*, 1508–1519. [CrossRef]
5. Pettersson, M. Silicon Nitride for Total Hip Replacements. Ph.D. Thesis, Uppsala University, Uppsala, Sweden, May 2015.
6. Berry, D.J.; Abdel, M.P.; Callaghan, J.J. What are the current clinical issues in wear and tribocorrosion? *Clin. Orthop. Relat. Res.* **2014**, *472*, 3659–3664. [CrossRef] [PubMed]
7. Kurtz, S.M.; Medel, F.J.; MacDonald, D.W.; Parvizi, J.; Kraay, M.J.; Rimnac, C.M. Reasons for revision of first-generation highly cross-linked polyethylenes. *J. Arthroplast.* **2010**, *25*, 67–74. [CrossRef] [PubMed]
8. Kurtz, S.M.; Gawel, H.A.; Patel, J.D. History and systematic review of wear and osteolysis outcomes for first-generation highly crosslinked polyethylene. *Clin. Orthop. Relat. Res.* **2011**, *469*, 2262–2277. [CrossRef] [PubMed]
9. Lee, J.; Salvati, E.; Betts, F.; DiCarlo, E.; Doty, S.; Bullough, P. Size of metallic and polyethylene debris particles in failed cemented total hip replacements. *J. Bone Jt. Surg. Br. Vol.* **1992**, *74*, 380–384. [CrossRef]
10. Drummond, J.; Tran, P.; Fary, C. Metal-on-metal hip arthroplasty: A review of adverse reactions and patient management. *J. Funct. Biomater.* **2015**, *6*, 486–499. [CrossRef]
11. Bitounis, D.; Pourchez, J.; Forest, V.; Boudard, D.; Cottier, M.; Klein, J.P. Detection and analysis of nanoparticles in patients: A critical review of the status quo of clinical nanotoxicology. *Biomaterials* **2016**, *76*, 302–312. [CrossRef]
12. Ayu, H.M.; Izman, S.; Daud, R.; Krishnamurthy, G.; Shah, A.; Tomadi, S.H.; Salwani, M.S. Surface modification on CoCrMo alloy to improve the adhesion strength of hydroxyapatite coating. *Procedia Eng.* **2017**, *184*, 399–408. [CrossRef]
13. Datta, S.; Das, M.; Krishna, V.; Bodhak, S.; Murugesan, V.K. Mechanical, wear, corrosion and biological properties of arc deposited titanium nitride coatings. *Surf. Coat. Technol.* **2018**, *344*, 214–222. [CrossRef]

14. Marchiori, G.; Lopomo, N.; Boi, M.; Berni, M.; Bianchi, M.; Gambardella, A.; Visani, A.; Russo, A.; Marcacci, M. Optimizing thickness of ceramic coatings on plastic components for orthopedic applications: A finite element analysis. *Mater. Sci. Eng. C* **2016**, *58*, 381–388. [CrossRef] [PubMed]

15. Grupp, T.M.; Giurea, A.; Miehlke, R.K.; Hintner, M.; Gaisser, M.; Schilling, C.; Schwiesau, J.; Kaddick, C. Biotribology of a new bearing material combination in a rotating hinge knee articulation. *Acta Biomater.* **2013**, *9*, 7054–7063. [CrossRef] [PubMed]

16. Ajwani, S.H.; Charalambous, C.P. Availability of total knee arthroplasty implants for metal hypersensitivity patients. *Knee Surg. Relat. Res.* **2016**, *28*, 312–318. [CrossRef] [PubMed]

17. Pettersson, M.; Tkachenko, S.; Schmidt, S.; Berlind, T.; Jacobson, S.; Hultman, L.; Engqvist, H.; Persson, C. Mechanical and tribological behavior of silicon nitride and silicon carbon nitride coatings for total joint replacements. *J. Mech. Behav. Biomed. Mater.* **2013**, *25*, 41–47. [CrossRef] [PubMed]

18. Mazzocchi, M.; Bellosi, A. On the possibility of silicon nitride as a ceramic for structural orthopaedic implants. Part I: Processing, microstructure, mechanical properties, cytotoxicity. *J. Mater. Sci. Mater. Med.* **2008**, *19*, 2881–2887. [CrossRef] [PubMed]

19. Bal, B.S.; Khandkar, A.; Lakshminarayanan, R.; Clarke, I.; Hoffman, A.A.; Rahaman, M.N. Fabrication and testing of silicon nitride bearings in total hip arthroplasty: Winner of the 2007 "HAP" PAUL award. *J. Arthroplast.* **2009**, *24*, 110–116. [CrossRef] [PubMed]

20. Rahaman, M.; Xiao, W. Silicon nitride bioceramics in healthcare. *Int. J. Appl. Ceram. Technol.* **2018**, *15*, 861–872. [CrossRef]

21. Pettersson, M.; Bryant, M.; Schmidt, S.; Engqvist, H.; Hall, R.M.; Neville, A.; Persson, C. Dissolution behaviour of silicon nitride coatings for joint replacements. *Mater. Sci. Eng. C* **2016**, *62*, 497–505. [CrossRef]

22. Pettersson, M.; Skjöldebrand, C.; Filho, L.; Engqvist, H.; Persson, C. Morphology and dissolution rate of wear debris from silicon nitride coatings. *ACS Biomater. Sci. Eng.* **2016**, *2*, 998–1004. [CrossRef]

23. Lal, S.; Hall, R.M.; Tipper, J.L. A novel method for isolation and recovery of ceramic nanoparticles and metal wear debris from serum lubricants at ultra-low wear rates. *Acta Biomater.* **2016**, *42*, 420–428. [CrossRef]

24. Hovsepian, P.E.; Ehiasarian, A.P.; Purandare, Y.P.; Biswas, B.; Pérez, F.J.; Lasanta, M.I.; De Miguel, M.T.; Illana, A.; Juez-Lorenzo, M.; Muelas, R.; et al. Performance of HIPIMS deposited CrN/NbN nanostructured coatings exposed to 650 °C in pure steam environment. *Mater. Chem. Phys.* **2016**, *179*, 110–119. [CrossRef]

25. Paulitsch, J.; Schenkel, M.; Zufraß, T.; Mayrhofer, P.H.; Münz, W.D. Structure and properties of high power impulse magnetron sputtering and DC magnetron sputtering CrN and TiN films deposited in an industrial scale unit. *Thin Solid Films* **2010**, *518*, 5558–5564. [CrossRef]

26. Ma, Q.; Li, L.; Xu, Y.; Gu, J.; Wang, L.; Xu, Y. Effect of bias voltage on TiAlSiN nanocomposite coatings deposited by HiPIMS. *Appl. Surf. Sci.* **2017**, *392*, 826–833. [CrossRef]

27. Bohlmark, J.; Lattemann, M.; Gudmundsson, J.T.; Ehiasarian, A.P.; Aranda Gonzalvo, Y.; Brenning, N.; Helmersson, U. The ion energy distributions and ion flux composition from a high power impulse magnetron sputtering discharge. *Thin Solid Films* **2006**, *515*, 1522–1526. [CrossRef]

28. Williams, S.; Isaac, G.; Hatto, P.; Stone, M.H.; Ingham, E.; Fisher, J. Comparative wear under different conditions of surface-engineered metal-on-metal bearings for total hip arthroplasty. *J. Arthroplast.* **2004**, *19*, 112–117. [CrossRef]

29. *ASTM F799-11 Standard Specification for Cobalt-28Chromium-6Molybdenum Alloy Forgings for Surgical Implants (UNS R31537, R31538, R31539)*; ASTM International: West Conshohocken, PA, USA, 2011.

30. *ISO 26443-2008 Fine Ceramics (Advanced Ceramics, Advanced Technical Ceramics) Rockwell Indentation Test for Evaluation of Adhesion of Ceramic Coatings*; ISO: Geneva, Switzerland, 2008.

31. *ASTM F2094/F2094M-18a Standard Specification for Silicon Nitride Bearing Balls*; ASTM International: West Conshohocken, PA, USA, 2018.

32. Olofsson, J.; Pettersson, M.; Teuscher, N.; Heilmann, A.; Larsson, K.; Grandfield, K.; Persson, C.; Jacobson, S.; Engqvist, H. Fabrication and evaluation of Si_xN_y coatings for total joint replacements. *J. Mater. Sci. Mater. Med.* **2012**, *23*, 1879–1889. [CrossRef] [PubMed]

33. *ASTM F1537-11 Standard Specification for Wrought Cobalt-28Chromium-6Molybdenum Alloys for Surgical Implants (UNS R31537, UNS R31538, and UNS R31539)*; ASTM International: West Conshohocken, PA, USA, 2011.

34. Fialho, J.C.; Fernandes, P.R.; Eça, L.; Folgado, J. Computational hip joint simulator for wear and heat generation. *J. Biomech.* **2007**, *40*, 2358–2366. [CrossRef]

35. Sanders, A.P.; Brannon, R.M. Assessment of the applicability of the Hertzian contact theory to edge-loaded prosthetic hip bearings. *J. Biomech.* **2011**, *44*, 2802–2808. [CrossRef]
36. *ASTM F732-17 Standard Test Method for Wear Testing of Polymeric Materials Used in Total Joint Prostheses*; ASTM International: West Conshohocken, PA, USA, 2017.
37. Archard, J.F. Contact and rubbing of flat surfaces. *J. Appl. Phys.* **1953**, *24*, 981–988. [CrossRef]
38. *ASTM F2033-12 Standard Specification for Total Hip Joint Prosthesis and Hip Endoprosthesis Bearing Surfaces Made of Metallic, Ceramic, and Polymeric Materials*; ASTM International: West Conshohocken, PA, USA, 2012.
39. Nedfors, N.; Mockute, A.; Palisaitis, J.; Persson, P.O.Å.; Näslund, L.Å.; Rosen, J. Influence of pulse frequency and bias on microstructure and mechanical properties of TiB_2 coatings deposited by high power impulse magnetron sputtering. *Surf. Coat. Technol.* **2016**, *304*, 203–210. [CrossRef]
40. Panjan, M.; Čekada, M.; Panjan, P.; Zupanič, F.; Kölker, W. Dependence of microstructure and hardness of TiAlN/VN hard coatings on the type of substrate rotation. *Vacuum* **2012**, *86*, 699–702. [CrossRef]
41. Ehiasarian, A.P.; Münz, W.D.; Hultman, L.; Helmersson, U.; Petrov, I.; Seitz, F. High power pulsed magnetron sputtered CrN_x films. *Galvanotechnik* **2003**, *163*, 267–272. [CrossRef]
42. Schmidt, S.; Hänninen, T.; Goyenola, C.; Wissting, J.; Jensen, J.; Hultman, L.; Goebbels, N.; Tobler, M.; Högberg, H. SiN_x coatings deposited by reactive high power impulse magnetron sputtering: Process parameters influencing the nitrogen content. *ACS Appl. Mater. Interfaces* **2016**, *8*, 20385–20395. [CrossRef] [PubMed]
43. Ortega-Saenz, J.A.; Hernandez-Rodriguez, M.A.L.; Ventura-Sobrevilla, V.; Michalczewski, R.; Smolik, J.; Szczerek, M. Tribological and corrosion testing of surface engineered surgical grade CoCrMo alloy. *Wear* **2011**, *271*, 2125–2131. [CrossRef]
44. Sahasrabudhe, H.; Bose, S.; Bandyopadhyay, A. Laser processed calcium phosphate reinforced CoCrMo for load-bearing applications: Processing and wear induced damage evaluation. *Acta Biomater.* **2018**, *66*, 118–128. [CrossRef]
45. Dante, R.C.; Kajdas, C.K. A review and a fundamental theory of silicon nitride tribochemistry. *Wear* **2012**, *288*, 27–38. [CrossRef]
46. López, A.; Filho, L.C.; Cogrel, M.; Engqvist, H.; Schmidt, S.; Högberg, H.; Persson, C. Morphology and adhesion of silicon nitride coatings upon soaking in fetal bovine serum. In Proceedings of the 15th International Symposium on Computer Methods in Biomechanics and Biomedical Engineering and the 3rd Conference on Imaging and Visualization, Lisbon, Portugal, 26–29 March 2018.
47. Saikko, V.; Calonius, O.; Kernen, J. Effect of counterface roughness on the wear of conventional and crosslinked ultrahigh molecular weight polyethylene studied with a multi-directional motion pin-on-disk device. *J. Biomed. Mater. Res.* **2001**, *57*, 506–512. [CrossRef]
48. Williams, S.; Tipper, J.L.; Ingham, E.; Stone, M.H.; Fisher, J. In vitro analysis of the wear, wear debris and biological activity of surface-engineered coatings for use in metal-on-metal total hip replacements. *Proc. Inst. Mech. Eng. Part H J. Eng. Med.* **2003**, *217*, 155–163. [CrossRef]
49. Leslie, I.J.; Williams, S.; Brown, C.; Anderson, J.; Isaac, G.; Hatto, P.; Ingham, E.; Fisher, J. Surface engineering: A low wearing solution for metal-on-metal hip surface replacements. *J. Biomed. Mater. Res. Part B Appl. Biomater.* **2009**, *90*, 558–565. [CrossRef]
50. Bal, B.S.; Rahaman, M.N. Orthopedic applications of silicon nitride ceramics. *Acta Biomater.* **2012**, *8*, 2889–2898. [CrossRef]
51. Olofsson, J.; Grehk, T.M.; Berlind, T.; Persson, C.; Jacobson, S.; Engqvist, H. Evaluation of silicon nitride as a wear resistant and resorbable alternative for total hip joint replacement. *Biomatter* **2012**, *2*, 94–102. [CrossRef] [PubMed]
52. Das, M.; Bhimani, K.; Balla, V.K. In vitro tribological and biocompatibility evaluation of sintered silicon nitride. *Mater. Lett.* **2018**, *212*, 130–133. [CrossRef]
53. Mak, M.; Jin, Z.; Fisher, J.; Stewart, T.D. Influence of acetabular cup rim design on the contact stress during edge loading in ceramic-on-ceramic hip prostheses. *J. Arthroplast.* **2011**, *26*, 131–136. [CrossRef] [PubMed]

MDPI

St. Alban-Anlage 66

4052 Basel

Switzerland

Tel. +41 61 683 77 34

Fax +41 61 302 89 18

www.mdpi.com

Coatings Editorial Office

E-mail: coatings@mdpi.com

www.mdpi.com/journal/coatings